The Ultimate Guide to Building a Google Cloud Foundation

A one-on-one tutorial with one of Google's top trainers

Patrick Haggerty

BIRMINGHAM—MUMBAI

The Ultimate Guide to Building a Google Cloud Foundation

Copyright © 2022 Packt Publishing

Group Product Manager: Rahul Nair

Publishing Product Manager: Niranjan Naikwadi

Senior Editor: Shazeen Iqbal

Content Development Editor: Romy Dias

Technical Editor: Rajat Sharma

Copy Editor: Safis Editing

Project Coordinator: Ashwin Dinesh Kharwa

Proofreader: Safis Editing

Indexer: Pratik Shirodkar

Production Designer: Prashant Ghare

Marketing Coordinator: Nimisha Dua

First published: July 2022

Production reference: 1220722

Published by Packt Publishing Ltd.

Livery Place

35 Livery Street

Birmingham

B3 2PB, UK.

ISBN 978-1-80324-085-5

www.packt.com

To my beautiful and loving wife Donna, who said, "Of course you should try and write a book," and then supported me through the months of nights and weekends it took to make that happen. Couldn't have done it without you, baby love.

Contributors

About the author

Patrick Haggerty was never quite sure what he wanted to be when he grew up, so he decided he'd just try things until he figured it out. Thrown out of college at 20, he spent 4 years in the USMC learning responsibility (and to be a better apex predator). Out on a disability, he turned wrenches in a mechanic shop, worked tech support, studied Actuarial Science, and coded in more languages than he wants to remember. When a job asked him to run some internal training, he discovered a lifelong passion: helping people learn.

Patrick has worked as a professional trainer for 25+ years and spends most of his days working for ROI Training and Google, helping people learn to leverage Google Cloud.

I'd like to thank Dave Carey, CEO of ROI Training (the #1 technical training org around), for being the best boss ever, and for keeping me solvent. I'd like to thank Packt for reaching out and encouraging me to write this book, and then for all the great people they brought in to help. Finally, I'd like to thank my fellow trainers and the people at Google who helped me answer all sorts of odd questions.

About the reviewer

Hector Parra worked in corporate IT for more than 15 years, specializing in Failure Monitoring and Automatic Recovery. Four years ago, he joined Google as a Customer Solutions Engineer, helping the biggest customers in Spain and EMEA to make the most out of Google Cloud for their marketing needs, whilst completing an executive MBA degree at Quantic. Hector is a certified Google Cloud Digital Leader and co-leads Google's Mind the Gap program in Spain, which was created to encourage more young women to pursue science and engineering careers. In his spare time, Hector is a big fan of retro gaming, TV shows, and electronic music. He loves traveling with his wife, Eva, and spending quality time with his big family, especially his two beloved nephews and five grandchildren.

I would like to thank my family for their patience with the time and effort required to review this book. My knowledge about the cloud wouldn't be the same without the amazing contribution of my colleagues at Google, both in the cloud and marketing areas, from whom I've learned so much. Thank you for these amazing last four years!

Table of Contents

6
Laying the Network

7
Foundational Monitoring and Logging

8
Augmenting Security and Registering for Support

Index

Other Books You May Enjoy

Preface

Only a few years ago, organizations were terrified of the cloud. These days, however, someone high up in the food chain goes to a conference and comes back saying *"We just HAVE to move into the cloud, NOW."*

If you start building IT services in Google Cloud, it won't take you long before you realize that there were things that you should have done before doing the things that you did do. The issue isn't that moving to the cloud is bad – it's more that there's a right way, and a hard way.

Do you remember when your parent said to you, *"I'm telling you this for your own good, so you won't make the mistakes I made."* Well, I've learned Google Cloud (mostly the hard way) and I wrote this book to help you use Google best practices to build a solid, secure, and extensible foundation on which you can set your IT services.

Who this book is for

If you've done a little experimenting in Google Cloud, but you can't quite decide how to start a real work environment that you and your organization can use as a foundation on which to build your IT services, then you've come to the right place. This book will teach you the parts of Google Cloud that are essential in any good foundation, and it will walk you through a best practices approach to building a solid, secure, and extensible foundation.

What this book covers

Chapter 1, Getting to Know Google's Cloud, is a short introduction and level set on Google Cloud, to make sure you enter the rest of the book with the base knowledge you need.

Chapter 2, IAM, Users, Groups, and Admin Access, introduces Google's 10 steps to building a solid foundation, and completes the first three: setup initial access to Google Cloud, configure the first users and security groups, and enable administrative access.

Chapter 3, Setting Up Billing and Cost Controls, will help you understand how Google Cloud charges, configure initial billing, and lay in budgets and alerts to help prevent cost overruns.

Chapter 4, Terraforming a Resource Hierarchy, introduces infrastructure automation with Terraform, and then use it to build an initial resource hierarchy, which will help with the application of security related policies.

Chapter 5, Controlling Access with IAM Roles, uses groups and Google predefined security roles to control permissions across our organization in Google Cloud.

Chapter 6, Laying the Network, builds a Virtual Private Cloud network to help isolate and secure network related resources.

Chapter 7, Foundational Monitoring and Logging, explains how to leverage Google Cloud instrumentation to better monitor and troubleshoot systems in Google Cloud.

Chapter 8, Augmenting Security and Registering for Support, extends and reinforces Google Cloud security measures, and how to register for support for when you need help from Google.

To get the most out of this book

A working knowledge of the terminal / command-line window is expected. Also, it would be helpful if you've played around in Google Cloud a little before reading the book.

Download the example files

You can download the example code files for this book from GitHub at `https://github.com/PacktPublishing/The-Ultimate-Guide-to-Building-a-Google-Cloud-Foundation`. If there's an update to the code, it will be updated in the GitHub repository.

We also have other code bundles from our rich catalog of books and videos available at `https://github.com/PacktPublishing/`. Check them out!

Download the color images

We also provide a PDF file that has color images of the screenshots and diagrams used in this book. You can download it here: `https://packt.link/FLbGs`.

Conventions used

There are a number of text conventions used throughout this book.

`Code in text`: Indicates code words in text, database table names, folder names, filenames, file extensions, pathnames, dummy URLs, user input, and Twitter handles. Here is an example: "`allAuthenticatedUsers` is a special placeholder representing all service accounts and Google accounts, in any organization (not just yours), including Gmail."

A block of code is set as follows:

```
resource "google_tags_tag_value" "c_value" {
    parent = "tagKeys/${google_tags_tag_key.c_key.name}"
    short_name = "true"
    description = "Project contains contracts."
}
```

When we wish to draw your attention to a particular part of a code block, the relevant lines or items are set in bold:

```
{
    "deniedPrincipals": [
        "principalSet://goog/public:all"
    ],
    "exceptionPrincipals": [
        "principalSet://goog/group/cool-role-admins@gcp.how"
    ],
```

Any command-line input or output is written as follows:

```
cd gcp-org
git checkout plan
```

Bold: Indicates a new term, an important word, or words that you see onscreen. For instance, words in menus or dialog boxes appear in **bold**. Here is an example: "If you click the **TROUBLESHOOT** button, Google Cloud will forward you to the Policy Troubleshooter."

> **Tips or important notes**
> Appear like this.

Get in touch

Feedback from our readers is always welcome.

General feedback: If you have questions about any aspect of this book, email us at customercare@packtpub.com and mention the book title in the subject of your message.

Errata: Although we have taken every care to ensure the accuracy of our content, mistakes do happen. If you have found a mistake in this book, we would be grateful if you would report this to us. Please visit www.packtpub.com/support/errata and fill in the form.

Piracy: If you come across any illegal copies of our works in any form on the internet, we would be grateful if you would provide us with the location address or website name. Please contact us at copyright@packt.com with a link to the material.

If you are interested in becoming an author: If there is a topic that you have expertise in and you are interested in either writing or contributing to a book, please visit authors.packtpub.com.

Share Your Thoughts

Once you've read *The Ultimate Guide to Building a Google Cloud Foundation*, we'd love to hear your thoughts! Scan the QR code below to go straight to the Amazon review page for this book and share your feedback.

https://packt.link/r/1803240857

Your review is important to us and the tech community and will help us make sure we're delivering excellent quality content.

1
Getting to Know Google's Cloud

I was consulting in Canary Warf, London, for a large financial services consultancy a few years ago when I got called into a meeting with a client who'd just signed a significant software development contract to be deployed into Google Cloud. I was invited to help answer Google Cloud questions. I arrived a little early wearing blue jeans, cowboy boots, and a T-shirt showing a father and son walking together. The son asks, *"Daddy, what are the clouds made of?"* The father replies, *"Linux servers mostly."*

Precisely on time, in strolled a British banker. *How did I know?* I'm American, and if an American closes their eyes and pictures a banker from a 150-year-old brick-and-mortar UK bank, yeah, *it was this guy*. Former RAF officer, three-piece suit, cufflinks. Suddenly, I felt way underdressed.

The meeting started, and Google Cloud was mentioned for the first time. Mr. Banker looked like he'd just eaten something sour and said (insert posh British accent), "*God, I hate the cloud.*" And, the room got really quiet. "*I hate the cloud, but as The Bank needs to stay competitive in the 21st century, we feel we have no other choice than to move some of our services into it. The cloud gives us the scalability, availability, and access to features that we need, so all the (nasty upstart) dot com start-ups don't get the best of us.*"

I tend to agree.

On the website for the US **National Institute of Standards and Technology** (**NIST**), they define **cloud computing** as: "*a model for enabling ubiquitous, convenient, on-demand network access to a shared pool of configurable computing resources (e.g., networks, servers, storage, applications, and services) that can be rapidly provisioned and released with minimal management effort or service provider interaction.*" (`https://csrc.nist.gov/publications/detail/sp/800-145/final`)

Yeah, that sounds like a government-created definition, doesn't it? But what does it mean in the context of Google Cloud?

In this chapter, we're going to answer this question by covering the following topics:

- How Google Cloud is a lot like a power company
- The four main ways of interacting with Google Cloud
- Organizing Google Cloud logically and physically
- Google's core services

Let's get started.

How Google Cloud is a lot like a power company

Most of you probably don't generate your own power. *Why?* It's an economy-of-scale thing. Your core business probably isn't power generation, so you likely don't know much about it and wouldn't do that good a job generating it if you tried. Besides, you and your business simply don't need enough power to make generating it yourself cost-effective. What do you do instead? You connect to a power grid, like this:

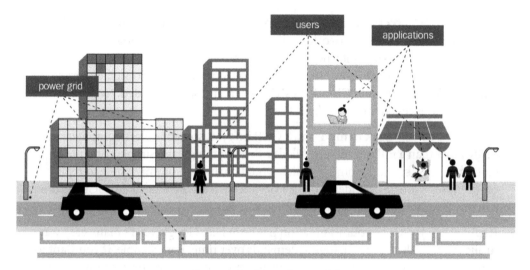

Figure 1.1 – The utility model

As you can see, the power company specializes in generating power at scale, and you plug in and use a little slice of that power however you need. Do you worry about how the power is generated? Not overly, but you do want the power to be there whenever and however you need.

Google Cloud Platform (**GCP**) works just like a power company, only for IT services. Instead of building your IT infrastructure from scratch, you tap into GCP and use what you need, when you need it.

I have another T-shirt that reads, "*There's no such thing as the cloud; it's just someone else's computers.*" It's funny, but it's also only part of the story. Circling back to that NIST cloud definition, it goes on to say that the cloud model has five key characteristics, all of which apply nicely to GCP:

- **On-demand self-service**: With Google Cloud, you can create your account, spin up and manage resources, monitor what those resources are doing, and shut down what you don't need, all from the comfort of your favorite browser, command line, or automation tool.

- **Broad network access**: Google has tens of thousands of miles of fiber comprising one of the largest private networks in the world. According to Google's estimates, more than 40% of global internet traffic now moves across Google's network. Within the global GCP network, you can carve out software-defined private networks known as **Virtual Private Clouds** (**VPCs**) when and where needed, to better control and regulate access to network-based services.

- **Resource pooling**: Back to my T-shirt, GCP isn't just someone else's computers. It's easily more than a million servers, filled with everything from simple compute to cutting-edge IT services, some of which would be difficult or impossible to replicate on-premises. Google provides those services at such a scale that again, GCP becomes a living, breathing example of an economy at scale.

- **Rapid elasticity**: Once your foundation has been laid, Google offers many easy ways to allow it to scale to meet changing demands over time. For example, if you want to construct a web application that can scale resources up and down based on load, GCP has the pieces and parts you'll need, from the load balancer distributing the incoming requests to the scalable pool of compute resources generating and assembling the responses, to the datastore behind it all.

- **Measured service**: Now, the bad news. I hosted a get to know Google Cloud session the other day and one of the attendees started a question: "*Does Google charge for…*" Let me stop you right there. Google measures your use of most everything, typically by the second, and bills you for it. But the news isn't all bad. If I go back to our scaling web application from the last point, while on-premises data centers are slow to scale because scaling involves adding or removing physical servers, the cloud scales very quickly. Since you pay for what you use, when your application scales down, you no longer pay for the services you aren't using. IT services become an operational rather than a capital expenditure. You rent what you need, exactly when you need it.

One of the cost-oriented business advantages of moving to GCP is that it can help provide the IT services you need at a lower overall **Total Cost of Ownership** (**TCO**). I would define TCO as everything you pay on your Google bill, added to anything you pay for stuff outside of, but related to, the systems you have in Google. As an example, take a Linux **Virtual Machine** (**VM**). I'm going to pay Google to host the VM, but if the VM is running Ubuntu Linux, then I'm going to need someone in my organization who understands the ins and outs of Ubuntu; how to secure it, harden it, patch it, and generally manage it, even if it isn't running in my on-premises data center. I'm also going to need know-how on securing around it, so who do I have who knows about network security as it relates to Google Cloud? Speaking of Google Cloud, do I know how to build a good foundation that I can drop my VM into? *I'm going to need some of those skills, too.* Does that mean I need to attend training? Buy a **book**? Yes! There also may be some costs just for integrating GCP into my organization. Do I need to upgrade my internet connectivity? Buy a VPN proxy? Run fiber to Google? Yeah, there may be other resource costs depending on those needs. So, *TCO = Google Cloud bill + external personnel + knowledge + resource costs.*

I'm getting a little ahead of myself here. For now, let's examine the various ways to interact with Google Cloud, and get it doing what we need.

The four main ways of interacting with Google Cloud

Imagine you've just visited `https://cloud.google.com/getting-started` and you now have a new account in Google Cloud. Maybe you even hit `https://workspace.google.com/` and created a new organization with a domain, email addresses, Google Drive, the works. What's the very first thing you should do next?

Wrong! What you should do next is build a foundation for you IT services to stand on in Google Cloud. When people build a house they don't just grab a board and throw it up anywhere. If you don't have a solid and secure foundation, whatever you put up is going to be unstable, unreliable, and insecure.

Now, if you want to know how to build that foundation, keep reading. We're going to get our journey started by examining the four major ways of interacting with GCP: through the Console, from the **Command-Line Interface** (**CLI**), through the APIs with code or automation software, and through the mobile app. Let's start with the Console.

Google Cloud Console

Google Cloud Console is the name Google uses for its web-based UI. On a side note, if your brain typically associates *console* with *command line*, then it's going to need to do some relearning when it comes to Google Cloud. You can get to the GCP Console via the link: `https://console.cloud.google.com/`. The Cloud Console overview page for a project looks as follows:

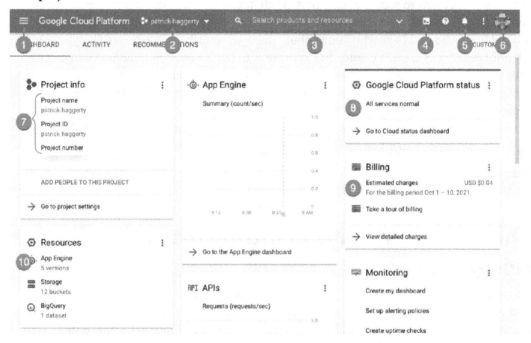

Figure 1.2 – Google Cloud Console home page

> **Note**
>
> Google Cloud is constantly evolving and constantly improving, which means little tweaks are made to the UI on a weekly, if not daily, basis. I work in GCP most days of most weeks, and I'm always coming across something new. Sometimes, it's "*Why did they move that?*" new, while others, it's more, "*What the heck?*" and it's time to investigate a new feature, new. This means that the moment I take a screenshot, such as the one shown in *Figure 1.2*, it starts to go out of date. Please forgive me if, by the time you read this, what you are seeing differs from this slice in GCP time.

In the preceding screenshot, I've labeled many of the key parts of the page, including the following:

1. The main Navigation menu hamburger button
2. The currently selected project
3. The auto-complete search box
4. Link to open Cloud Shell
5. Alerts and messages
6. Current user avatar
7. Information about the currently selected project
8. Google Cloud's current status
9. Billing information
10. List of key resources

Here, we can see the home page of my `patrick-haggerty` project.

> **Note**
>
> Using **Personally Identifiable Information** (**PII**) when you're creating projects and folders is not a good idea. Do as I say, not as I do.

We will formally talk about projects in the next section, so for now, a project is a logical group of resources with a billing account attached and paying for them all. I know the project I'm in because the project drop menu (*2*), in *Figure 1.2*, tells me so. That drop menu will appear on any project resource page I navigate from here, lest I forget or think I'm in a different project. I can also see which account I'm currently logged in under (*6*) because I am using a different avatar image for each of my accounts. If I need to switch accounts, I can do it there.

The project home page contains a lot of nice information. There's an overview of the project-identifying information (*7*), a list of any known current issues in Google Cloud (*8*), a summary of my bill for the month so far (*9*), and a list of key resources I've created in the project (*10*).

While the project home page is just a single page in GCP, the top title bar is standard and appears on most pages. When you want to move between the various Google Cloud products and services, you can either use the Navigation menu (*1*) or the search box (*3*). Did you notice my use of title case for the Navigation menu? I'm going to do that so that you know when I'm talking about that hamburger menu (*1*).

Now, I teach Google Cloud classes multiple times every month, and I am in the habit of using the Navigation menu because I like to show students where things are. I also have a terrible memory for details and sometimes, browsing through the product list helps remind me what I'm looking for. When I was new to GCP, it also showed me the available products. As a result, I rarely use search. That's not typical. I've watched lots of people work with GCP and a lot of them prefer the search box.

So, if I wanted to launch BigQuery and go run some SQL against my data warehouse, I would go to the Navigation menu, then **ANALYTICS**, and click on **BigQuery**, as shown in the following screenshot:

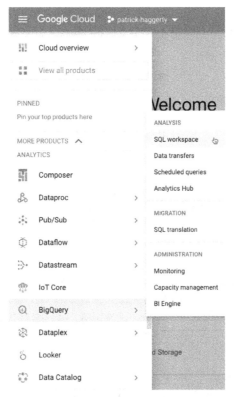

Figure 1.3 – Finding BigQuery in the Navigation menu

By selecting **BigQuery**, it defaults to that first item in the **SQL workspace** fly-out menu, but I could have selected anything in the submenu as required.

If you're visually impaired and not driving with the mouse, or if you just love keyboard commands, pressing *?* at any time will bring up a list of keyboard shortcuts. There, you will see that pressing *.* will open the main Navigation menu, where you can move around with arrow keys and select your choice with *Return/Enter*. Here's a screenshot showing the keyboard shortcuts:

Keyboard shortcuts

Action	Shortcut
Open products and services	.
Go up one page	u
Open project navigator	Cmd + o
Find products and services	/
Open shortcut help	? or Cmd + Shift + /
Send feedback	@
Open help menu	g then h
See all notifications/activity	g then n
Activate Google Cloud Shell	g then s

CLOSE

Figure 1.4 – Keyboard shortcuts

If you're more of an auto-complete search box sort of user, or if you are new and find that driving by menu is exhausting, then you're going to love the search box (back in *Figure 1.2*). Google knows a thing or two about search (*snort*). Entering text not only works for product names, but it will also find projects, resources you've created with that name, and it will even suggest documentation where you can find more information. So, you could also find BigQuery via the search menu by clicking into the search box or pressing / and typing what you're looking for, like so:

Figure 1.5 – BigQuery by search

I tend to interact with GCP through the Console when I'm initially building things and I'm in the *not sure what I need so I'm experimenting* phase. Mostly, though, I find that the UI is the absolute best way to monitor the things I have running.

Driving by UI is nice, but sometimes, you just need a command line.

The Google Cloud SDK and Cloud Shell

If you want to do some simple automation from the CLI, or if you just really love the command line, then you'll like **Cloud Shell** and the **Cloud SDK**. The Cloud SDK is a downloadable set of command-line utilities. I have a section on this near the top of my helpful links file (`http://gcp.help`). First, you must download and install the SDK (`https://cloud.google.com/sdk/docs/install`). Once you've installed it, open your terminal and type the following:

```
gcloud init
```

You will be walked through authenticating into your Google Cloud account, selecting a default project, and (optionally) a default region and zone (more on that in the next section). Once the setup is complete, you can interact with Google Cloud from the command line on your laptop. There are several CLI utilities available, but the most commonly used two are as follows:

- gcloud: gcloud is a CLI way of saying, *"Hey Google, here's what I need you to do."* The order of arguments goes from most general, through the verb, to specific options. So, if I wanted to get a list of **Google Compute Engine** (**GCE**) VMs, then I'd execute the following command:

  ```
  gcloud compute instances list
  ```

  I'd read that as, *"Hey gcloud." "Google here, with whom would you like to interact?"* *"Compute." "Good, but Compute Engine is a huge product, so what part of Compute Engine?" "Instances." "Oh, I got you. Do you want to create an instance?" "No, give me a list."* In this case, I didn't need to specify any options, so that part has been left out.

  For more information, see `https://cloud.google.com/sdk/docs/cheatsheet` and `https://cloud.google.com/sdk/gcloud/reference`.

- gsutil: We'll get into some specifics on products in Google Cloud later in this chapter, but suffice it to say that **Google Cloud Storage** (**GCS**) for offsite file storage is a biggie, and gsutil is the CLI tool for interacting with it. Its arguments are reminiscent of Linux file commands, only prefaced with gsutil. So, if I wanted to copy a file from my local system up into my patrick-haggerty Cloud Storage bucket, the command would look as follows:

  ```
  gsutil cp delete-me.txt gs://patrick-haggerty
  ```

  For a reference, see `https://cloud.google.com/storage/docs/gsutil`.

Besides downloading and installing the Google Cloud SDK, there is another way to use the CLI, and that's through Google's web-based **Cloud Shell**. If you recall my GCP web page interface overview back in *Figure 1.2*, number **4** was just below the link for opening Cloud Shell. When you open it for the first time, Cloud Shell will appear as a terminal window across the bottom of the web page, sharing space with the GCP Console:

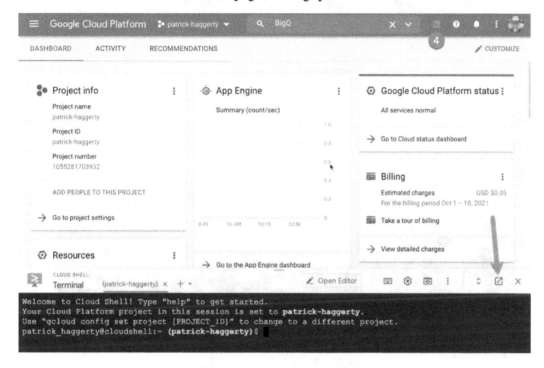

Figure 1.6 – Opening Cloud Shell

Now, I don't particularly like the Google Cloud Console sharing space with the Cloud Shell terminal, so I like to click that up-to-the-right icon and open Cloud Shell in a separate tab. Once the new tab comes up and stabilizes, you can close the Cloud Shell terminal at the bottom of the console page. As a result, you'll have one browser tab displaying the GCP Console and another displaying Cloud Shell.

Here, you can see I've done exactly that. My first tab is displaying the Console, while the second tab (which you can see in the following screenshot) is showing Cloud Shell with both the editor and the terminal active:

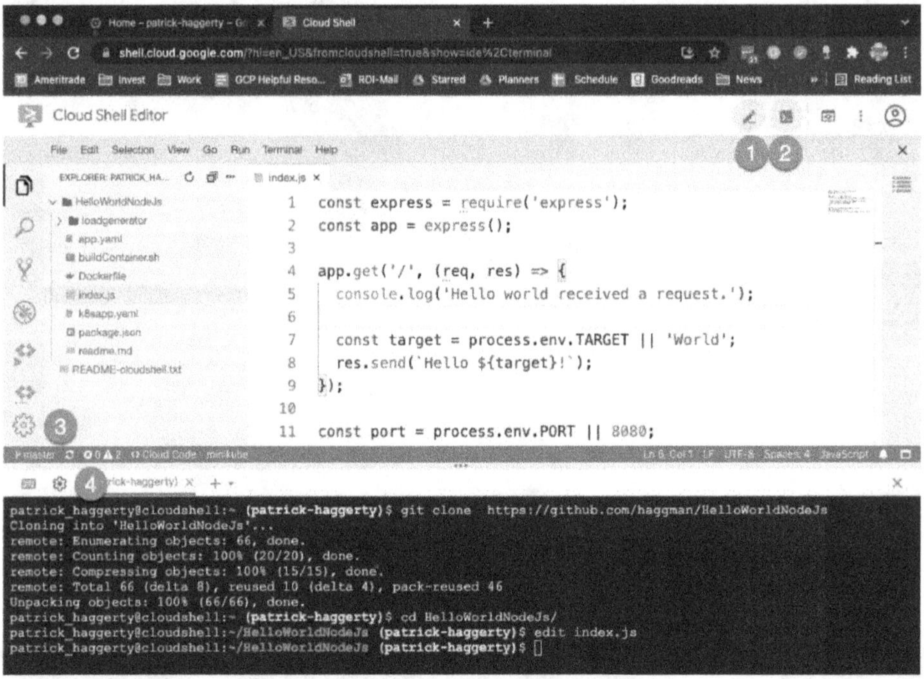

Figure 1.7 – Cloud Shell on a separate tab

In the preceding screenshot, I've labeled several key features:

1. Toggle on/off the Theia editor

2. Toggle on/off the terminal window

3. Theia editor settings (font size is a handy one)

4. Terminal settings

What Google does to enable Cloud Shell is launch a (free) virtual machine, drop in your user's hard drive, and give you the interface you can see in the preceding screenshot. The Cloud Shell hard drive persists across sessions and is tied to your user account. It isn't very large (5 GB), but you can store a file on it now and it will still be there when you log in next time, provided it's within the next 120 days. Google also has all the SDK CLI tools installed, along with all sorts of Linux utilities, coding languages and APIs, and they even have **Theia** installed. Theia (besides being a Titan and sister, who is also his wife (I know, gross), to Hyperion) is part of the Eclipse editor family, but it looks a lot like **Visual Studio Code** (**VS Code**) and supports many VS Code features and extensions.

Cloud Shell can be a big help when you need to pop into GCP, edit a configuration file, and push a quick change. The CLI in general is frequently the first step toward basic automation. Need to automate the creation of a couple of VMs so that you can do it over and over? How about a simple Bash script with a couple of `gcloud` commands? That will get the job done, but some automation software, such as Terraform, may do a better job.

The Google Cloud APIs

Everything in Google Cloud is a service. Service can mean *"something that provides value"* and GCP services are that, but each GCP service is also a web API service. If you know how to format the correct JSON input, and you send it the right URL, authenticated the right way, then you could use GCP that way too.

To see an example, you can head over to Google's APIs Explorer at `https://developers.google.com/apis-explorer`. Locate the API for Compute Engine and with a little hunting around, you'll see that there's an `instances.list` function. Here's a direct link: `https://cloud.google.com/compute/docs/reference/rest/beta/instances/list`. Over on the side is a **Try It!** button. The API explorer will prompt you for input parameters, and you'll have to enter the name of a project and a compute zone where the VMs live. Once you've done that, click **Execute** and it will send a JSON request to GCP, as well as showing you the input and returned message, as shown here:

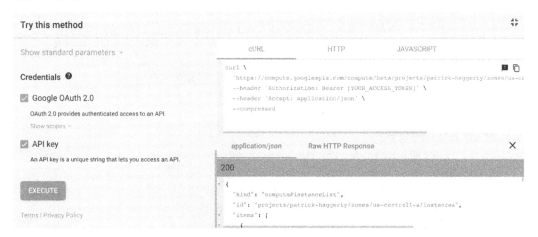

Figure 1.8 – Manual API request

This is a nice example, but this isn't the way to use the APIs – not if you are working in one of the languages Google has created client APIs for. Take a look at `https://cloud.google.com/apis/docs/cloud-client-libraries`. To use that same command, get a list of VMs in my `patrick-haggerty` project, which is situated in `us-central1-a`, while working in *Node.js*:

```
const compute = require('@google-cloud/compute');
const projectId = 'patrick-haggerty';
const zone = 'us-central1-a'
async function quickStart() {
    const instancesClient = new compute.InstancesClient();

    const [instanceList] = await instancesClient.list({
      project: projectId,
      zone,
    });
    for (const instance of instanceList) {
        console.log(`${instance.name}`);
    }}
quickStart();
```

Now, besides custom code accessing Google Cloud services, the API is also the way the various automation tools, such as Terraform, interact with Google. We will talk about Terraform a bit later.

For now, what if you're on the phone and you need to check in with Google Cloud?

The Google Cloud mobile client

Yes, Google Cloud has a client you can run on your mobile device. Yes, it is a quick and easy way you can investigate what's happening in Google Cloud. No, it is not the preferred way of interacting with GCP in a general sense. If you want to overview what's happening in the cloud, or perhaps do some quick monitoring, then it's a good tool. You can also send alerts to individual mobile devices, so it may help with alerting. Here, you can see a couple of screenshots of my mobile app, where I chose to look at the list of VMs I have running in my `patrick-haggerty` project:

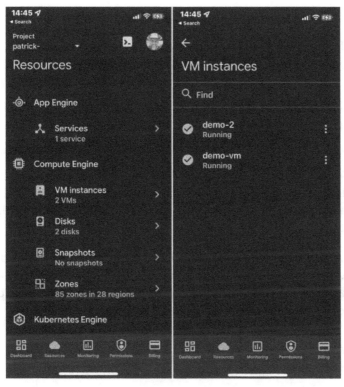

Figure 1.9 – The Google mobile app

Honestly, my favorite feature on the model app is that little Cloud Shell link in the top-right corner. Typically, if I wanted to see the resources I had in my project from my iPhone, I'd just use the web page. The standard GCP Console works well on phones and tablets. But the web page version of Cloud Shell on my iPhone? *Not so fun*. The mobile app, which has a custom keyboard, is the only way to use Cloud Shell on a phone.

Great – now that you're clicking, typing, coding, or tapping your way into Google Cloud, how exactly does Google structure resources?

Organizing Google Cloud logically and physically

Everything in Google Cloud has both a *logical* and a *physical* organization to it. A virtual machine, for example, could logically belong to a department and physically run in London, and decisions need to be made around both these organizational forms. Let's start with the logical aspect.

Definition time! The core elements of Google Cloud's logical organization are as follows:

- **Resource**: In that NIST definition they stated, "*a shared pool of configurable computing resources (e.g., networks, servers, storage, applications, and services)*" so likely, if it's part of Google Cloud and costs money, then it's a resource.

- **Project**: A logical collection of resources with an owner and with a billing account (paying for all the resources in this project) attached.

- **Folder**: A logical collection of projects and/or other folders that's used to create groupings to form logical or trust-related boundaries.

- **Organization**: The root node of the logical organization hierarchy. It's tied to an organizational domain, has some form of identity, and has an organizational administrator.

Chapter 4, Terraforming a Resource Hierarchy, of this book will go into some details of setting this logical structure up. For now, let's look at an example. If you have access to view your organizational structure, then you can go to the Navigation menu, click on **IAM & Admin**, followed by **Manage resources**. You will see something like this:

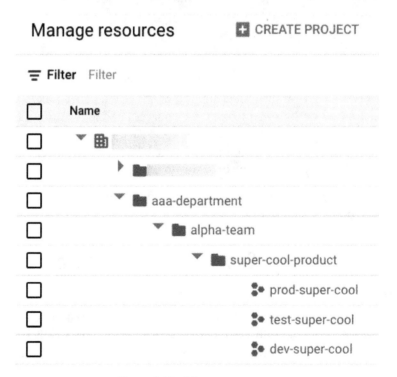

Figure 1.10 – Manage resources

Here, you can see that I've blurred out my organization name and that top-level folder, but under that is the structure I created for this example. I have a folder for `aaa-department`, which creates a logical unit of organization and a trust boundary. Inside that folder, I have another folder for the `alpha-team` developers. Under that, I have created three different projects for my super-cool application – one where we do development work, one where we do the testing and staging, and one where we have super-cool running in production. I'm not saying this is the way you must do things, but this is an example of a hierarchy I've seen used in production.

If the organization, folder, and project are the major units of logical organization, how does physical organization work?

Physically, Google Cloud is broken down into data centers, where we can create resources. The key definitions that are related to physical organization are as follows:

- **Region**: A data center in a particular geographic area of the world. `europe-west2` is in London, England.

- **Zone**: A unit of isolation inside a region. With a single exception, all generally available data centers have three isolation zones. `europe-west2-a`, `b`, and `c` would be the sub-units of the London region.

- **Multi-Regional**: Some GCP services can run in two or more regions simultaneously, typically to offer better redundancy and higher availability.

- **Global**: A service that is available and running in every GCP data center in the world.

Let's start with an example. `us-central1` is one of the largest Google Cloud regions, and the only region in the world to have four zones. It's located in Council Bluffs, Iowa, and you can find it on Google Maps if you search for it. It looks like this:

Figure 1.11 – Google Cloud us-central1 data center

Do you know what that is? It's a big concrete box full of hardware, with lots of security, and very few people. Inside that big box are four zones (remember, most regions have three), which I would imagine as smaller concrete sub-boxes. Each zone is essentially a failure domain. If `us-central1-a` fails, that failure is unlikely to impact b, c, or f (*yeah, don't ask*). In the entire history of Google Cloud, they've never lost a whole region, but they've had zonal failures.

Speaking physically, every product in Google Cloud is realized in a zone, in multiple zones of a region, in multiple regions, or globally. When you're planning an architecture, one of the things that should be on your to-do list is to determine where exactly to run each of the services you're planning on using. Only by understanding a GCP service's inherent redundancy, and your needs, will you be able to plan your regions.

Speaking of GCP services and how they are logically and physically organized, how about we look at a few concrete examples in the next section?

Google's core services

Google Cloud is simply enormous, both in terms of its physical size and features. A nice place to see this enormity is the **Google Cloud Developer's Cheat Sheet** (`https://github.com/priyankavergadia/google-cloud-4-words`). There, you will find a list of most Google Cloud services, each one with a short description and a pair of links where you can find more information. The chain-link icons take you to the product's overview page, and the links that resemble dog-eared pages take you to the product's main documentation page.

The Google Cloud Developer's Cheat Sheet also has an excellent service overview graphic:

Figure 1.12 – Google Cloud Developer's Cheat Sheet

Time to get out the magnifying glass! It probably works better as a poster, but you get the idea. They even now have a dynamic version you can find here: `https://googlecloudcheatsheet.withgoogle.com/`

A big part of moving to the cloud is using the right products, the right way. There is a term people toss around: **cloud native**. The first time I heard about cloud native, it concerned moving code to the cloud. If you write or rewrite your application to best leverage what the cloud offers, then it becomes cloud native. Over the last few years, however, the term has taken on a life of its own and now it tends to mean the general concept of best utilizing what the cloud offers, not just in terms of coding.

Now, this book is attempting to help you lay a solid Google Cloud foundation, which means I'm not going to be doing a detailed analysis of a lot of Google Cloud's services. This also means that you will need to spend time digging into service details before you use them. And you may also wish to keep an eye out for (*shameless plug*) some other books that Packt and I have in the planning stage.

There's a lot of ways you can categorize Google's services, and we can only comfortably examine the major ones, so let's focus on the two most ubiquitous categories: **compute** and **data**. If you want more information on any product I mention, see the links in the Developer's Cheat Sheet. I also have a constantly evolving **Google Helpful Links Doc**: `http://gcp.help`. If your organization doesn't like my redirect URL, or if you can't access Google Docs at work, try at home or on your mobile device and create a PDF copy. Just remember that I am always adding/tweaking/evolving the live document, so your PDF will slowly become out of date. If you prefer, I've created a PDF copy of my own here: `http://files.gcp.how/links.pdf`, which I update every couple of weeks or so.

One final note before I dig into some individual services: everything you do in the cloud becomes part of a **shared responsibility model**. Whichever product you use and whatever it does for you, you will own and control parts of it, and Google will own and control other parts. Always make sure you research exactly what you're responsible for (securing, configuring, managing, and so on), and what Google moves off your plate and takes care of for you.

Now, let's look at some services, starting with *compute*.

Compute

If you run software in the cloud, whether it's custom coded by your organization or something you buy off the shelf, then it's going to likely run in one of Google's compute services.

Now, there's a **Developer Advocate** who works for Google named Priyanka Vergadia who calls herself the *Cloud Girl*. You can find a lot of good videos on Google Cloud that have been created by her on the Google YouTube channel (`https://www.youtube.com/user/googlecloudplatform`), and she also has a website (`https://thecloudgirl.dev/`) where she has created a lot of fantastic graphics related to Google Cloud. She has one apropos to our current discussion:

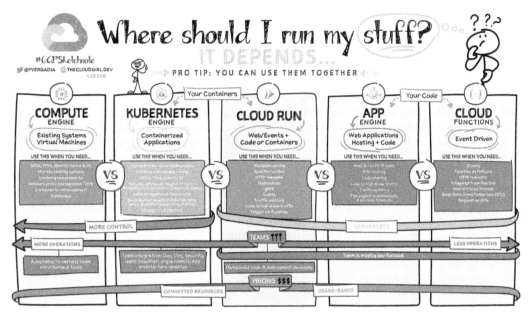

Figure 1.13 – Where should I run my stuff?

I'm going to approach the GCP compute technologies in the same order as the Cloud Girl by moving from left to right, starting with *GCE*.

Google Compute Engine

GCE is someone else's (very secure, very reliable, feature-rich) computer. If you are used to ordering servers, or you regularly spin up VMs using technologies such as vSphere, KVM, or Hyper-V, and you want to do virtually (cough) the same thing in Google Cloud, then this is the product you're looking for. You can pick the chipset, the vCPU count, the memory, the size of the boot disk, what OS comes loaded on it, whether it has externally accessible IPs, drive type sizes and counts, and a whole slew of other things. Then, you press the button, and in no time, your server will be up, running, and accessible.

If you need a group of identical VMs all cloned from the same base image, then GCE also offers **Instance Groups**. An Instance Group joined up with one of GCP's load balancers is a good way to build highly scalable, highly available, Compute Engine-based applications.

Of all the Google compute options, this is the one that's most frequently utilized by organizations that are new to GCP. It's a VM, and most organizations know how to create and manage VMs. They already have the Linux, Windows, networking, and security experts in-house, precisely because it's a type of compute they are used to working with. The problem? It's frequently the least cloud-native compute option with the highest TCO, so using it for every possible workload is not a best practice.

The following are the pros of GCE:

- Of all the GCP compute options, this is the ultimate in terms of flexibility and configurability.
- It's familiar to organizations currently using on-premises VMs.
- It's the single most popular first compute product for companies migrating into GCP.
- It's easy to scale vertically (bigger machine) and/or horizontally (more machine clones).
- Pay per second with a number of price break options for long running machines.

The following are the cons of GCE:

- All that flexibility means more to manage, control, and secure.
- Frequently requires specialized skillsets, just like on-premises VMs.
- It has a high relative TCO and you take a big slice of the shared responsibility pie.

Use GCE when you're doing the following:

- Lifting and shifting VMs to the cloud with as few changes as possible.
- Running custom or off-the-shelf workloads with specific OS requirements.
- You really need the flexibility provided by a VM.

The following are the location options you have:

- **Zonal**: A VM, or an Instance Group with multiple VMs, in a single zone.
- **Regional**: A VM can never span more than a single zone, but an Instance Group could place VM clones in different zones across a region.

Now, instead of having a full, *do-it-yourself* VM, if your application is containerized and you need a good container orchestration system, then you'll love *Google Kubernetes Engine*.

Google Kubernetes Engine

Google Kubernetes Engine (**GKE**) is a fully managed implementation of the wildly popular container orchestration technology, **Kubernetes** (also known as **k8s**, because there are 8 letters between the **k** and the **s**). If you want an overview of what a container is, check out `https://www.docker.com/resources/what-container`. If you want to see how Kubernetes can help you manage containers at scale, then look at the *What is Kubernetes?* page over on the main Kubernetes site (`https://kubernetes.io/docs/concepts/overview/what-is-kubernetes/`) and spend a few minutes watching (not a joke) *The Illustrated Children's Guide to Kubernetes* on YouTube (`https://youtu.be/4ht22ReBjno`).

Quick side discussion: in the preceding paragraph, I introduced a term you hear a lot regarding GCP: **fully managed**. Fully managed means that Google is going to take care of the implementation details for you. If you wanted to stand Kubernetes up on-premises, you might use something such as **kOps** to help with the installation, and you would be responsible for setting up the machines and managing them, the OSs, the OS and k8s patches, the security – the works. Instead, if you spin up a cluster in GKE, Google takes over the responsibility of managing the OS, the machines, and their security and patching. Google will even update your k8s version if you select a **release channel** option. Essentially, Google hands you an up-and-running set of machines, all ready and waiting with k8s installed – all you have to do is use it. GKE is one of Google Cloud's fully managed-by-Google services.

GKE has two operational modes: **Standard** and **Autopilot**. Standard mode is mostly what I described previously. You select your options, then Google builds your k8s cluster and hands you the keys. From there, Google keeps k8s up, running, up to date, and happy, and you use it however you want.

The following are the pros of GKE Standard:

- Easy way to get a turnkey Kubernetes cluster in under 5 minutes.
- Managed by Google with a typically lower TCO than GCE.
- Based on a widely used open source technology.
- Very popular with companies modernizing toward containers.
- Offers discounted and committed-use pricing for long-running clusters.

The following are the cons of GKE Standard:

- If your application isn't containerized, then you can't run it in k8s.
- Like any k8s cluster, you require specialized k8s skills to manage it.
- Not as flexible as GCE.

Use GKE Standard when you're doing the following:

- Running containerized workloads and you need control over the k8s environment.
- The container is long-running, stateful, or accessed through non-HTTP protocols (Cloud Run is not an option).
- You have in-house k8s skills, and you don't want to pay for Autopilot.

The following are the location options you have:

- **Zonal**: All cluster VMs in a single zone.
- **Regional**: A group of machines in each of three zones of a given region.

Besides GKE Standard, Google also offers an Autopilot version, where Google does more managing and you focus more on your workloads. But first, there's a new term I need to introduce: **serverless**. When Google talks about serverless products, what they mean is that you don't see, configure, or in any way manage the infrastructure where the product runs.

It kind of sounds like fully managed, *right?* But it's not quite the same. If you construct a GKE Standard cluster, there are actual VMs out there and you can see them if you look in GCE, but where a standard Compute Engine VM is managed by you and your team, in a fully managed situation such as GKE Standard, the machine is there but Google is taking care of it and its needs. Google then bills you a nominal cluster management fee, plus the cost of the VMs in your cluster. If your cluster is busy or doing almost nothing, it doesn't matter in GKE Standard because you're paying for the cluster itself.

GKE Autopilot is serverless, so Google takes on all the responsibilities related to Kubernetes administration and configuration, all the way down to selecting how many machines of what type comprise the cluster. Essentially, there's a k8s cluster out there, but the shared responsibility line has moved, and Google is taking on more of the administrative tasks for it.

I'll admit, the term *serverless* makes my eye twitch when I think about it too much. Of course there are servers out there somewhere just like with a GKE Standard cluster, but the difference here is they are now abstracted to such a degree that I no longer even see them as part of my project.

The following are the pros of GKE Autopilot:

- Essentially, this is the Kubernetes easy button.
- The k8s cluster becomes Google's problem.
- You create your containers, submit them to the cluster, and they just work.
- GKE Autopilot also offers discounted and committed-use pricing for long-running pods.

The following are the cons of GKE Autopilot:

- Not as flexible in terms of supported workloads.
- You have less control over cluster configuration (a double-edged sword).
- Pod resources are not as configurable.
- You pay a standard GKE cluster management fee, and a per second fee for the resources utilized by each of your pods.

Use GKE Autopilot when you need to run a container that can't run in Cloud Run, and you don't have the in-house k8s skills to go with GKE Standard.

There is a **Regional** location option.

For a good Autopilot and Standard GKE cluster comparison, see `https://cloud.google.com/kubernetes-engine/docs/concepts/autopilot-overview#comparison`.

OK; so, go for GKE Standard if you want to run a container and you need full control over the Kubernetes environment, but use GKE Autopilot if you want to run a container and you're willing to pay Google to manage the Kubernetes environment for you. But what if you want something cheaper and even easier than Autopilot? Oh, then you'll love *Cloud Run*.

Cloud Run

Cloud Run is another fully managed serverless (*see what I did there?*) technology for running containers. Where GKE Autopilot is a k8s cluster that's managed completely by Google, in Cloud Run, there is no k8s cluster. Google has constructed some tech for running containers in a super-flexible, super-easy-to-use, super-scalable environment of their devising (originally built for App Engine). Then, they combined that with an open source add-on for Kubernetes called **Knative**, which simplifies how developers configure and deploy particular types of containers. Pulling in Knative helps avoid vendor lock-in by providing an API that can be used to deploy and manage applications in Cloud Run, just as it could with any normal Kubernetes cluster running Knative.

OK – I know what you're thinking. You're wondering why Cloud Run exists when it sounds mostly like the same product as GKE Autopilot. This isn't going to be the only time you feel that way, trust me. Frequently, Google Cloud offers several different ways to do the same thing. I mean, we're in the middle of talking about compute technologies, right? All of them do the same thing in general, but they have strengths and weaknesses, and the devil is in those details. Truly being cloud native in GCP is going to require patience and research.

Let's get back to Cloud Run.

Cloud Run specializes in **stateless** (don't remember user data between requests) containers, which listen and serve requests on a configured port, doesn't run outside serving requests, are short-lived (return an answer in less than 60 minutes), and use the HTTP/1 or HTTP/2 (including gRPC) protocols. In other words, if you're building a web application or a web service API, and that page or API takes a request, does something, returns an answer, and then moves on to serve the next request, then this is a great place to run it.

Cloud Run also supports the **CloudEvents** open source standard, so instead of running in response to an HTTP request, a container could automatically react to something else going on in Google Cloud. For example, let's say that a new file just showed up over here in Cloud Storage. You could wake up a Cloud Run container, have it translate the file from English into Spanish, store a Spanish text file, and then go back to sleep.

Cloud Run also has the option to **scale to zero**. Scale to zero, when applied to a GCP service, means that if no one is using the service, then it will scale down automatically until it costs you nothing. That's right – if you aren't using Cloud Run, you can configure it such that you won't be paying for Cloud Run. In GKE Autopilot, so long as you have deployed a pod, you will be paying for the pod per second. In Cloud Run, Google measures the time it takes to handle a particular client request, rounds it to the closest 100 ms, and bills you for that. No requests? Nothing to bill for.

Cloud Run also has a **free tier**. That is, if your usage is below a certain threshold, then it doesn't cost you anything. Cloud Run is one of more than 20 products in GCP that has free tiers (`https://cloud.google.com/free/docs/gcp-free-tier`). I could have mentioned this in the GKE discussion, because the GKE cluster management fee for a zonal or Autopilot cluster, which is currently $.10/hour, is free for your first cluster, though you're still paying by VM or by pod. In Cloud Run, though, it's *free-free*. Your Cloud Run container runtime has to reach a base level of use before it costs you anything. While you are coding and testing, and until you hit that baseload level, you can get going with an app in Cloud Run for free.

The following are the pros of Cloud Run:

- Super-easy way to run HTTP1/2-based web and web service containers.
- Super-easy way to create automated cloud event handlers.
- It has a short and sweet learning curve with no Kubernetes to learn or manage.
- Auto-scales and auto-load balances.
- Managed implementation of open source Knative and CloudEvents projects.

The following are the cons of Cloud Run:

- Containers must be in Linux x86_64 ABI format, short-lived, stateless, and be accessed via requests coming in through HTTP1/2 or via a predefined set of triggers.
- Limited configurability.

Use Cloud Run when the workload is a container hosting a short-lived, stateless web app or a CloudEvents (trigger), and you don't need any of the configurability GKE provides.

The location option you have is Regional, where each service will be served out of a single GCP region, though you can run multiple clone Cloud Run instances in different regions and then load balance across them.

As cool as Cloud Run is, it wasn't the first GCP product to work this way. Long before Cloud Run was a gleam in the eye of the Google Cloud engineering team, there was *App Engine*.

App Engine

App Engine is one of the original Google Cloud (2008) services. It was cool long before Kubernetes, before Docker, and before the term container was a byword in the development community. Behind the scenes, it operates essentially the same way and in the same servers as Cloud Run. Arguing that Cloud Run is a next-generation replacement for App Engine wouldn't be far off the mark. Also, like Cloud Run, it primarily supports HTTP-accessible requests, which means more web applications and more web service API endpoints.

In Cloud Run, you build your container, put it in a registry somewhere Cloud Run can access, create your service, and you're off to the races. With App Engine, you are restricted to particular versions of particular languages: Java, Node.js, Python, C#, PHP, Ruby, or Go. You upload your source code and a configuration file to App Engine using a command-line tool, and Google creates a sandbox around your application and runs it in that sandbox.

App Engine offers multiple ways to scale, so it can be completely automatic like Cloud Run, or you can do basic or manual scaling where you have a lot more control to allow it to do things such as background tasks.

The following are the pros of App Engine:

- Easy to deploy and use, with a small learning curve
- Another option for web applications and API endpoints
- Can auto-scale and auto-load balance
- Scales to zero

The following are the cons of App Engine:

- Proprietary to Google, with an App Engine-specific config file.
- Runs in a sandbox, so your code has to be particular versions of particular languages.
- Some of the original App Engine support services have been moved out to make them more generally available.
- You should probably be looking at Cloud Run.

The location option you have is Regional, where each service will be served out of a single GCP region.

App Engine isn't designed to support events, such as CloudEvents triggers in Cloud Run, but another member of the App Engine family can. Let's find out what that is.

Cloud Functions

Cloud Functions is event-driven functions that run in essentially the same place as Cloud Run containers and App Engine applications. The function can respond to internal cloud events, such as a new file appearing in a Cloud Storage bucket or a new message showing up in a Pub/Sub topic, or it can act as a basic endpoint responding to HTTP requests. Like App Engine, you create Cloud Functions by coding them in particular versions of essentially the same list of languages. At the time of writing, Cloud Functions is being extended to also support CloudEvents APIs such as Cloud Run Triggers.

The following are the pros of Cloud Functions:

- You can easily create event handlers for several common GCP event types.
- Cost-effective (to a point) and scales to zero.

The following are the cons of Cloud Functions:

- The list of event types is limited (though CloudEvents will extend).
- Proprietary to Google (again, not if you're using CloudEvents).
- It uses sandboxes, so your code has to be particular versions of particular languages.
- Consider Cloud Run and triggers as a more container friendly alternative.

The location option you have is Regional, where each service will be served out of a single GCP region.

Great – so we've looked at several popular Google Cloud compute services, but what's compute without data? I'm sure that most applications that run in or out of the cloud depend on some sort of data storage, so let's turn around and take a peek at what Google offers in the way of data storage.

Data storage

Compute, data storage, and data processing were probably the three prime driving forces behind Google creating their cloud. Google currently has nine products with over a billion users each: Android, Chrome, Gmail, Google Drive, Google Maps, Google Search, Google Play Store, YouTube, and Google Photos. And all those users? They upload 400 hours or more of videos to YouTube (a petabyte) every minute. They upload 18 PBs of pictures to Google Photos per day. And that's not even talking about the huge storage and processing problem that is Google Search.

Cloud Girl time again. The way Google currently categorizes data storage products is into storage and databases. Let's start with **storage**:

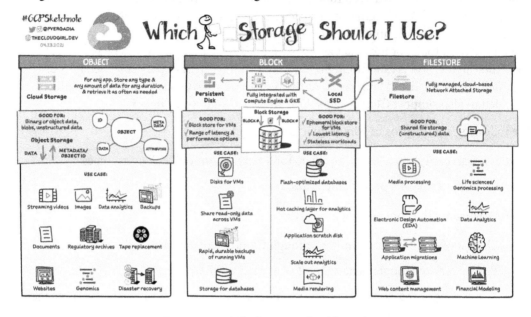

Figure 1.14 – Which storage should I use?

Let's take a closer look at Google Cloud's storage options.

Google Cloud Storage

GCS is a **Binary Large Object** (**BLOB**) store and, together with Compute Engine, it is one of the most popular first-step Google Cloud products. It's designed to be a cheap and easy way to store files off-site. Of all Google Cloud's data products, and probably all the GCP products period, it's something that most people are familiar with. Even non-technical people these days have stuff backing up to Google Drive, Google Photos, iCloud, and the like.

GCS sits on an enormous storage array that Google has constructed, called **Colossus**. You can find some information online about Colossus in the Data section of my `http://gcp.help` document. One of the features that Colossus passes to GCS is 99.999999999% (11 9s) of durability. This means that the chance of you loading a file into GCS and of Google then losing it is next to nothing. To put this into perspective, the chance of you not winning the typical lottery is about 99.99999% (7 9s), which is about a tenth of the chance of being bitten by a shark. You're way more likely to be struck by lightning, with a chance of about 99.999% (5 9s). *You get the idea.*

The following are the pros of GCS:

- Easy to use, highly durable, and always encrypted
- Supports versioning, locks, auto-deletes, and automated life cycle configurations
- Little to configure and you don't have to tell Google whether you're storing a little or a lot
- Low cost and you only pay for the actual data that's stored
- Low latency, with a time to the first byte in the tens of milliseconds
- Could store a virtually unlimited amount of data
- Several different pricing options for active and inactive data storage

The following are the cons of GCS:

- Does not provide block storage.
- Folders are simulated.
- Append operations and file moves necessitate full file replacements.
- It's probably not good for operations that require speed over many, many, small files.

Use GCS in the following cases:

- You want a cost-effective, off-site file store.
- You need to host images, `.js`, `.css`, or other static files behind a web application.
- You need stage files for batch data processing.

The following are the location options you have:

- **Regional**: Files are replicated across all the zones of a single region.
- **Multi-regional**: Files are copied across all the zones of two regions, at least 100 miles apart, and in the same region of the world (such as the US).
- **Dual-regional**: Same as multi-regional but you know the exact two regions.

Well, that's one way to store files, but what if you want something that's block storage and more like a traditional hard drive?

Persistent disks

Persistent Disks (**PDs**) are used when you need durable block storage. If you are running a GCE VM or a GKE cluster that needs drive space, then this is where that data lives. They aren't physical disks in the traditional sense. Instead, they are block storage backed by Colossus, so you still get the same 11 9s of durability and auto-encryption.

PDs are subdivided into several subtypes, depending on the IOPS, throughput, and spend requirements, and there's even a special PD type called Local SSDs, which is optimized for when you need blazing IOPS performance but only for temporary cache storage.

The following are the pros of PDs:

- Designed for persistent block storage – think laptop hard drives
- Resizable on demand (up, not down)
- Automatic encryption with several options as to how the keys are managed
- 11 9s of durability and excellent overall performance
- Easy to back up and restore with geo-replicated snapshots

The following are the cons of PDs:

- You pay for what you allocate, regardless of use.
- If you just want simple file storage, consider Cloud Storage.
- Restricted to a single zone, or pair of zones.

Use PDs in the following circumstances:

- When you need a drive for a VM or GKE workload
- When your application requires block storage

The following are the location options you have:

- **Zonal**: All the file data is stored within a single zone.
- **Dual-zonal**: There's an active drive in one zone, while a passive duplicate is kept in a second zone in the same region.

What if you need something more along the lines of a traditional, on-premises, network shared drive? For this, there's *Filestore*.

Filestore

Filestore (not to be confused with Firestore) is a fully managed **Network-Attached Storage** (**NAS**). It supports access from any NFSv3-compatible client and has several different configurable options, depending on your price, IOPS, throughput, and availability goals. GCE VMs or GKE workloads that need to share a read/write file storage location could leverage Filestore.

The following are the pros of Filestore:

- Fast (IOPS and throughput) and easy-to-use file store
- Predictable pricing
- Supports shared read and write access
- Networked

The following are the cons of Filestore:

- NAS is a file store rather than a block store.
- You pay for what you allocate, which can be expensive.

Use Filestore in the following situations:

- When your workload requires a NAS
- When you need to share read/write access to a drive

The following are the location options you have:

- **Zonal**: When set to *Basic* or *High Scale tier*
- **Regional**: When set to *Enterprise tier*

When your storage needs lean away from a simple object or file storage and toward databases, Google has options for you there as well:

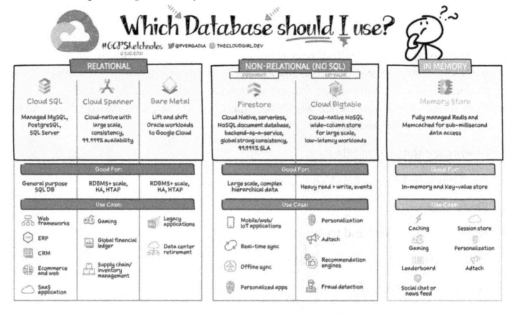

Figure 1.15 – Which database should I use?

Databases and **Database Management Systems (DBMSs)** are designed to aid applications by offering fast and secure storage and information retrieval. Having said that, the preceding diagram shows multiple options because, within that purview, databases tend to be optimized for particular usage models. Let's start with the classic relational options in the following two sections.

Cloud SQL

Relational databases have been a mainstay in the database world for decades. They are flexible, efficient, and do a good job with transactional data. If you want a classic relational MySQL, PostgreSQL, or Microsoft SQL Server database, and you'd like Google to set it up and manage it for you, then go no further than **Cloud SQL**.

The following are the pros of Cloud SQL:

- Classic relational database backed by MySQL, PostgreSQL, or MS SQL Server.

- The machine, DBMS, and backups are all managed and maintained by Google.

- You can configure the read/write instance size to fit your load and budgetary requirements (bigger machine, can do more work, but costs more money).

- Optional HA configuration with an active copy in one zone, and a passive backup in another, with auto-failover.
- Can scale out read-only replicas.

The following are the cons of Cloud SQL:

- The single read/write machine means scaling writes happens vertically. The machine can get bigger in terms of resources, but there are limits.
- Writes will always be made to a single machine, in a single zone.
- Maintenance events, when Google is patching or updating the OS and/or DBMS, can take the database down for upward of 60 seconds (check out the **Maintenance window** settings).

Use Cloud SQL in the following circumstances:

- You need a small to midsized, zonal, relational database.
- You're migrating on-premises MySQL, PostgreSQL, or MS SQL Server databases to GCP, and would like to make the system a little more cloud native.

The following are the location options you have:

- **Zonal**: The single active read/write instance will always be in a single zone.
- **Regional or multi-regional**: Only applied to the read-only replicas.

I like MySQL, PostgreSQL, and MS SQL Server as much as the next guy, but what about when I need a mission-critical relational database and its write requirements are either too large for what even the biggest single machine can handle, or it needs to be geographically distributed for latency reasons? Well, in that case, you should take a closer look at *Spanner*.

Spanner

Spanner is a Google-invented relational database. Its origin goes back to the early 2010s and a need to be able to store advertising-related data with strongly consistent transactions at a level bigger than a single region. Spanner is a strongly consistent, transactional, relational database that can horizontally scale to meet almost any demand.

As I sit here writing this, I'm wearing a pair of Lucchese cowboy boots. Lucchese boots are beautiful, comfortable, and you can wear them all day long, but you're going to pay for the privilege.

Spanner is the relational database equivalent to my boots.

The following are the pros of Spanner:

- Strongly consistent, transactional, and relational.
- Linear scaling with processing units (100 at a time), practically without boundaries.
- Serverless with no maintenance windows.
- A database can be bigger than a single region.
- Offers an on-premises simulator for development.
- You get exactly what you pay for.
- Can enable support for the PostgreSQL ecosystem.

The following are the cons of Spanner:

- It is not simply a managed version of an existing DBMS, so you will need someone to learn to administer and optimize it.
- Scaling up processing units, especially across regions, can get expensive.

Use Spanner in the following circumstances:

- You need a relational database that can scale horizontally for large loads.
- You need a single relational database that can operate across multiple regions.
- Your workload can't tolerate the Cloud SQL maintenance cycle.

The following are the location options you have:

- **Regional**
- **Multi-regional**: Within the same continent and multi-continent options

Relational databases may be a classic, but what if you're willing to move past the traditional and into the realm of non-relational storage? A good option to consider here would be *Firestore*.

Firestore (Datastore)

Firestore (not to be confused with Filestore) is a document-style NoSQL database. NoSQL (terrible name) means non-relational. Firestore doesn't store data as tables of records of fields. Instead, it's a document or entity database, and it stores *"things"* as chunks of data. If you've ever worked with a JSON store such as CouchDB or an object store such as MongoDB, then this is the Google equivalent.

Think about an order. An order is a hierarchal structure that likely contains some order details such as date and order total, but it also contains some customer data such as shipping and billing address, and a list of order items with counts and product details. We think "*order*" because it's a logical unit. In a relational database, you'd shred that data into different tables and do a join when you want the whole order back again. With Firebase, you can skip the whole shred/join steps because document-style databases can natively store complex structures such as orders, products, reviews, or customer states.

The following are the pros of Firestore:

- Fully managed non-relational document store database.
- Strongly consistent and transactional.
- Cost-effective, with a free tier and can scale to zero.
- Easy-to-use API and (non-SQL) query language.
- Scales fast and well.

The following are the cons of Firestore:

- Not all developers are familiar with NoSQL datastores.
- No SQL, no join queries, and no aggregations.

Use Firestore in the following circumstances:

- When you are storing "*things*" and a document/entity store can do the job.
- You are developing web or mobile devices and need an easy, cost-effective database.

The following are the location options you have:

- **Regional**
- **Multi-regional**

If I'm willing to go NoSQL, are there any databases that are built for sheer performance and throughput? Yes – if you want a datastore you can firehose data into for something such as IoT sensor data purposes, then check out *Bigtable*.

Bigtable

Bigtable is exactly that – it is a wide-column NoSQL datastore consisting of a single big table (similar to HBase or Cassandra). How big? That's up to you, but there could be hundreds or even thousands of columns and a virtually unlimited number of rows. This is the datastore behind Google products such as Search and Maps. Bigtable doesn't support transactions in the traditional sense, and by default, it is eventually consistent, so changes in one cluster will eventually replicate everywhere, but this can take seconds to minutes. The only indexed column is the row key, which means that schema design is extremely important, especially in regards to the row key. It is another GCP product in the *"get what you pay for"* category, so make sure it's the right option before you choose it.

The following are the pros of Bigtable:

- Managed, wide-column, NoSQL datastore.
- Ability to scale to handle virtually any required R/W performance.
- Extremely low latency.
- Supports the HBase API.

The following are the cons of Bigtable:

- You will have to learn how to configure and use it.
- Not transactional and eventually consistent.
- It can get pricey.

Use Bigtable in the following circumstances:

- When you are working with IoT, Adtech, fast lookup, or time series data and you need the ability to read and/or write many records with extreme performance.
- Store the data here, because it's so fast, then analyze it later, with code or other tools.

The following are the location options you have:

- **Regional**
- **Multi-regional**

If you're looking for a highly available memory cache, rather than a brute-force performance database, then you should look into the two options provided by *Memorystore*.

Memorystore

Memorystore is a highly available, Google-managed version of the open source memory caches **Redis** or **Memcached**. It is designed to act as a sub-millisecond, in-memory cache. So, it's another NoSQL database, but this time, it's a simple key-value store. You pick the open source cache engine and Google takes it from there.

The following are the pros of Memorystore:

- It provides a sub-millisecond in-memory cache that's 100% compatible with OSS Redis or Memcached.
- Fully managed, highly available, and secure.
- Different service tiers and sizes, depending on your need.

The following are the cons of Memorystore:

- Some features from the OSS versions are not supported.
- It has a simple key-value store that's optimized for fast lookup.

Use Memorystore in the following circumstances:

- When you need a shared super-fast, in-memory key-value store
- When you're caching session information, a shared set of products, a leaderboard, and more

The following are the location options you have:

- **Zonal**
- **Regional** (cross-zone active-passive replication with automatic failover)

Whew – that's a lot of services and, honestly, we've barely scratched the surface. What you need, though, before you start throwing services at a problem, is a secure and scalable foundation to construct them on. That's right – and you've come to just the right place to figure out how.

Summary

In this chapter, we started by providing a high-level overview of the cloud itself, likening it to a power company, and illustrating some of its key advantages: on-demand, broad network access, resource pooling, elasticity, pay for what you use, and economy of scale. From there, we moved on to examine the four ways of interacting with Google Cloud: through the Google Cloud Console (web UI), the command line (SDK and Cloud Shell), the APIs with code or automation software such as Terraform, and via the mobile app. Lastly, we did a quick pass over several different Google Cloud services in the two key areas: compute and data.

From here, there are a lot of directions we could go, but this is a book about laying a Google Cloud foundation, so we need to keep focused on that. Well, if you want to lay a good foundation, then you should probably start with the next chapter.

2
IAM, Users, Groups, and Admin Access

Building IT services for your organization in **Google Cloud** needs to start like a good house, with a firm and secure foundation. If the foundation is weak, then sooner or later, cracks appear, and that's so much harder to fix after the house is built.

Interestingly, this is exactly what got me into working with Google Cloud. I've spent most of my career writing code, helping others to write code, training around writing code, and… well, you get the picture. I've been coding in one form or another since the mid-1980s. I never wanted to be a cloud consultant/trainer; I liked the *code*.

A Note on the Whole "Why Google Cloud?" Question

I regularly get asked why I like Google Cloud. There are a lot of good reasons, but they are based on my experiences and the types of work I do. I say the best answer is, *"Go try it yourself."* AWS was the first and is currently the biggest cloud provider, followed by Azure (rhymes with *pressure*), with Google Cloud the third most popular. All the major cloud providers have free trials and other programs to allow you to give things a spin before you spend too much money. Do some tutorials, try to do what you need, and see what you like best. If it's Google Cloud, then you're reading the right book.

As a developer, I followed my aforementioned advice and did some work with AWS, a bit with Google Cloud, and working with SharePoint got me into a bit of Azure, but after playing with the big three cloud providers, the one that clicked best for me was Google. So, I started saying to my clients, *"Hey, this app you want to build – why don't we build it in Google Cloud?"*

Those who were at least open to the idea came back with one of two common responses. First, there were the *"Sounds like an interesting idea… How do we do that?"* clients. They liked the idea of the cloud and were open to giving it a shot, but they didn't know how to make a start. Understandably, they wanted help taking those first steps, and there I was.

Now, I am not a good liar, and if I don't just say what I'm thinking, it typically shows all over my face, so for the clients who went with *"Great idea – as a matter of fact, we've already started moving to Google Cloud. The CEO/CTO/CIO/big boss recently attended a conference, and we've started an initiative to create all future applications in the cloud. Come look,"* it was really hard to not let the horror show.

Why horror? Well, because so many times, I would look at what they had done and it would be hard not to simply put my head in my hands. There would be so much fundamental cloud architecture to fix, and that was all before we could even start with the application development.

It's generally not a good idea to burst out with uncontrollable laughter or to tell a client that they're an idiot, so I usually went with something like, *"Yes, I can see what you've done there and I'm glad you're thinking cloud, but there's a couple of things we should probably rework before building the application."* Hey, embellishment isn't lying; it's just good storytelling.

At some point, my brain kicked in and said, *"Hey, there's a need here. You should do this,"* and then I wasn't writing as much code anymore.

To help get our foundation started in **Google Cloud Platform** (**GCP**), Google has created a checklist, which you can find here: `https://cloud.google.com/docs/enterprise/setup-checklist`. I'm not going to reinvent the wheel by coming up with my own checklist, but I am going to paraphrase Google's a bit:

1. Configuring identity management
2. Adding an initial set of users and security groups
3. Enabling administrator access
4. Setting up billing and initial cost controls
5. Creating a resource hierarchy to control logical organization
6. Adding IAM trust boundaries to the resource hierarchy
7. Building and configuring the initial **Virtual Private Cloud** (**VPC**) network

8. Configuring logging and monitoring so that you know what's happening in the cloud

9. Adding organization policies, the Security Command Center, and other security measures

10. Selecting and enabling a Google support plan

At the time of writing, Google is creating a wizard to help you through their steps (https://console.cloud.google.com/cloud-setup). Most of this help seems to be taking the form of instructions with links, then Google runs checks to see that you've completed each item, and you mark the items complete once you're both satisfied. I'm going to walk you through the steps in detail so that you know what you're doing and why. Feel free to just use the help that the wizard provides, but keep reading to truly understand the decisions that you and the wizard are making.

In this chapter, we're going to take the first three major steps toward laying our foundation in Google Cloud:

- Step 1 – configuring identity management

- Step 2 – adding an initial set of users and security groups

- Step 3 – enabling administrator access

Step 1 – configuring identity management

I tossed around a few different ideas on how to get this section started, and I finally realized that the easiest would be to simply set up a new organization that I could use for any examples related to this book. I headed over to a **Domain Name System** (**DNS**) registry service (I like https://www.namecheap.com/) and did a little unused domain name searching. I thought this had a nice ring to it:

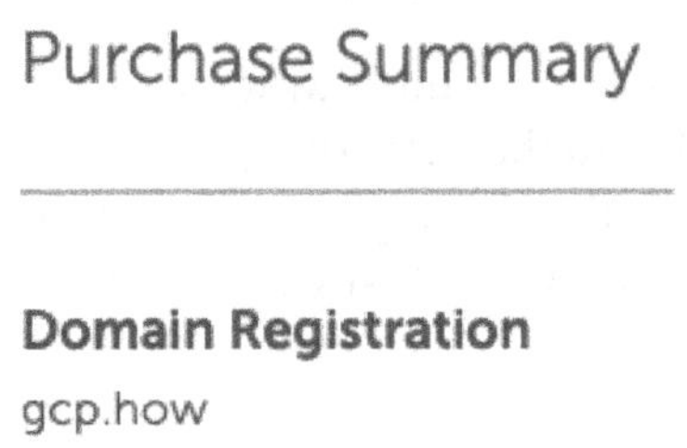

Purchase Summary

Domain Registration
gcp.how

Figure 2.1 – My new domain – gcp.how

Then, I signed up for **Google Workspace** (formerly **G Suite**). To be clear, having a Google Workspace is not a requirement for setting up Google Cloud. I'm using Google Workspace because it provides me with email, Google Drive, and so on, but it doesn't automatically come with an account in Google Cloud.

Google workspaces can be self-served from the Workspace home page, `https://workspace.google.com/`, or a domain service such as **Namecheap** can also create the workspace account for you.

Now that I have my domain and email configured, let's get back to our first major Google Cloud setup step – **configuring identity management**.

Cloud Identity is Google's **Identity as a Service (IDaaS)** offering. IDaaS is a cool acronym for a system that can authenticate users. If you head into an office building to visit someone, you frequently have to show your ID before you can get access. That's an analog version of an IDaaS system. Google's Cloud Identity can handle identity services for you directly, as it both stores and validates credentials, or it can federate with other providers, such as Microsoft's **Active Directory (AD)** and Azure AD, and let them be the source of truth when it comes to identity.

In the case of Google Cloud, Cloud Identity will be the access gateway that decides exactly how apps, people, and devices prove who they are in a secure and reliable way before they can access services. But Cloud Identity can be much more than just an IDaaS system; it can also help organizations manage employee computers, phones, and tablets, as well as integrating into systems requiring **Lightweight Directory Access Protocol (LDAP)**.

In my office building example, after the person at reception checks your ID to make sure you are who you claim, they then have to decide on exactly what you can access. In Google Cloud, once identity is confirmed, **Identity and Access Management (IAM)** takes over and, based on the security roles your identity has been assigned, IAM can then decide exactly what you are authorized to do in terms of Google Cloud services. We will talk more in depth about IAM in a later chapter.

Together, Cloud Identity and Google Cloud IAM work like *Figure 2.2*. The user comes into Cloud Identity and proves who they are. Then, based on the roles directly assigned to them or to the group to which they belong, they are authorized to access some, all, or none of the resources that Google Cloud makes available:

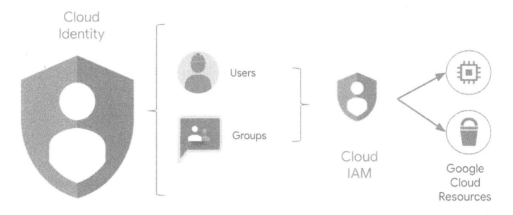

Figure 2.2 – Cloud Identity working with Cloud IAM

Before setting up Cloud Identity, you need to decide which of the two editions you'll need – **free** or **premium**. If you look at `https://cloud.google.com/identity/docs/editions`, you'll see a good comparison. The editions aren't mutually exclusive, so you can allocate some of your users to free and others to premium, based on need.

To briefly summarize the differences, the free version provides the authentication you'll need for Google Cloud for up to 50* users, plus Chrome sync, most Cloud Identity reporting features, and core mobile device and computer management. The premium edition (currently *$6 a month per user*) provides everything from the free pile for as many users as you want to pay for, plus much more advanced mobile device and mobile app management, LDAP app integration support, more reporting, and 24 x 7 email, phone, and chat support, all with a 99.9% SLA.

For my demonstration site, where I'm mostly concerned with controlling access to Google Cloud, the free version will be just fine.

---

*Up to 50 Users

The default Cloud Identity free user count is officially 50 but, based on your organization's reputation and what else you may be paying for, Google can automatically and/or manually raise that number. For example, I'm using Cloud Identity's free version, but my default user count limit in Cloud Identity is showing 90 users. Why? Because I'm paying for Google Workspace, so I'm already paying something per user.

Cloud Identity setup

To set up Cloud Identity, you will need the following:

- The ability to get into your domain name registry settings so that you can verify domain ownership. If you have any domain aliases (perhaps you own .com and .org), have the keys to them ready as well.

- A first Cloud Identity super admin username.

I have just set up my domain, so I know that I have access, but you'd be surprised how frequently this is a stumbling point for organizations. Everyone in the organization knows that there is a domain name, and you pay for it every 5 years or something, but who has the keys to it? Check with IT and, if they don't know, send a message to your finance department and have them check for the bill. If no one has a clue, you can do an **Internet Corporation for Assigned Names and Numbers (ICANN) lookup** (https://lookup.icann.org/). I just did a search for my new domain name, and it provided me with a contact form that I can use to send a message to the owner. It also told me who the registrar was and how I could contact them. Going down that route may take a bit of time, but it's a place to start.

Next, I need to decide on my first Cloud Identity super admin user. This will be the first user who has full control over Cloud Identity, and that's a lot of power at the access control level. The best practices for Cloud Identity super admins are as follows:

- The account should be used for Cloud Identity administration (user and group life cycle and organizational security settings) only. It should not be a normal daily-use account.

- The account should not be used as a Google Cloud organizational administrator (Google Cloud's version of a super admin, which we'll discuss in more detail later).

- You should eventually have two to five Cloud Identity super admins. You need at least two in case one is on vacation. You don't need many more than five because it's just too much power to have floating around.

- This first account will be used for initial Cloud Identity or Google Workspace setup, Google Cloud setup, and then kept back for use as a final recovery option. If you are a super admin, we'll create an account for that later, but not yet.

- As a best practice, come up with a descriptive but generic name, such as `gcp-superadmin@<your-domain>`. I'm going to use `gcp-superadmin@gcp.how`.

- Protect all super admin accounts with some form of **2-Step Verification (2SV)**. The strongest option would be a physical key, such as Google's own Titan key (`https://cloud.google.com/titan-security-key`), or one of the keys from Yubico (`https://www.yubico.com/products/`). Make sure that you also have backup configured, perhaps backup codes encrypted and stored securely.

You can find the steps to sign up for Cloud Identity here: `https://cloud.google.com/identity/docs/setup`.

If you aren't going to be pairing Cloud Identity with Google Workspace, start with one of these:

- To sign up for Cloud Identity Free, start here: `https://workspace.google.com/signup/gcpidentity`.

- To sign up for Cloud Identity Premium, start here: `https://cloud.google.com/identity/signup/premium/welcome`.

If you are going to use Google Workspace, then set that up first and after it's up and running, follow these steps:

1. Log into your Google Workspace admin page: `https://admin.google.com/`.
2. Go to **Billing | Get more services | Cloud Identity**.
3. Select **Cloud Identity Free** (my choice) or **Cloud Identity Premium**.

The steps you run through to initially sign up for Google Workspace or to sign up for Cloud Identity are almost identical. There's a setup wizard, and you'll have a quick *about you* section where, besides your name, you'll also have to provide a secondary email, something outside your current domain. There's a section where you can enter *basic business information*, your *domain information*, and finally, you'll finish and run through the initial user setup. Remember, this should be your first super admin user, as discussed previously, so it should be your equivalent to my `gcp-superadmin@gcp.how`.

The *domain name verification* section is where you'll need to be able to access your domain name registrar configurations. In it, you'll have to create a **TXT** or **CNAME** record, depending on preferences, or you may also be able to verify your domain by adding a tag or page to your organizational website itself. Since I also needed to tie my domain name mail configurations to my Google Workspace, I had to add a handful of **Mail Exchanger** (**MX**) records. It can take Google minutes to hours to complete its verification process once the registrar configurations are updated. You may have to check back on the verification pages multiple times. For me, it took about 10 minutes.

Once that part is complete, you'll get your Cloud Identity/Google Workspace admin page: `https://admin.google.com/`.

Now that you have initial access to Cloud Identity or your Google Workspace, via the corresponding superuser, go ahead and log into Google Cloud with it for the first time by heading over to the getting started page for Google Cloud: `https://console.cloud.google.com/getting-started`. If you like, go ahead and start your trial, though you'll need to be ready to provide a credit card if you do (make it a corporate card!).

If you can access the Cloud Identity or Google Workspace admin page (`https://admin.google.com`) with your superuser account, and you can access the Google Cloud Console page (`https://console.cloud.google.com/`), then you are ready to move on to the next major step.

Step 2 – adding an initial set of users and security groups

Before we get to adding an initial set of users, we need to discuss how you manage users now and whether you want to integrate that with what you're doing in Cloud Identity.

There are two main concepts that come into play here:

- The one source of truth for user identities. This is where you create and manage users and where you store information about those users (such as name, email, manager, and group membership).
- The **Identity Provider** (**IdP**) where you perform authentication to verify that a user is who they claim to be.

So, the big questions are, what are you going to use as the one source of truth for users? What are you going to use as an IdP? And if they aren't the same, how are you going to mix them?

Cloud Identity managing users and acting as IdP

Let's start with the case where Google Cloud Identity is managing both users and serving as an IdP, something like this:

Cloud Identity managing both users and acting as IdP

Figure 2.3 – Cloud Identity managing both users and the IdP

The nice thing about this approach is that it's easy, low-cost, and you don't have to do anything to integrate Google Cloud Identity with an external IdP or user service. You are essentially setting up a new and isolated set of users, and you're using a Google mission-critical security system to do it.

This is when you should use it:

- You're just getting started as a new business, and leveraging Cloud Identity as both a user store and IdP just makes sense. In this case, I'd strongly consider Google Workspace as well, since it does such a nice job with Docs, Sheets, Drive, and so on.

- You are already using Google Workspace and simply want to add Google Cloud access.

- You want a small, isolated set of users to have access to Google Cloud services and you like the idea that they are managed independently in GCP.

This is what the login process will look like to a user:

1. The user follows a link to a protected resource in GCP, such as Cloud Storage.

2. If their browser isn't already authenticated, then they are redirected to a standard Google sign-in screen where they enter their username and password. If 2SV is enabled, they are prompted to provide their key or code.

3. If the user authenticates, then IAM decides whether they are authorized to access the requested resource, and if so, they are then redirected back to the service they requested – in our example, Cloud Storage.

That's really easy, but what if an organization likes Google as an IdP but they are already managing users with a system in HR? In that case, a better option might be to split the duties.

Cloud Identity managing IdP and an HR system managing users

Some organizations use a **Human Resources Information System (HRIS)** such as SAP SuccessFactors, BambooHR, Namely, Ultimate Software, or Workday to manage their users. By letting the HRIS focus on user management, you can preserve existing HR workflows for things such as provisioning new users, and Cloud Identity can be used to handle everything IdP-related. This architecture would look something like this:

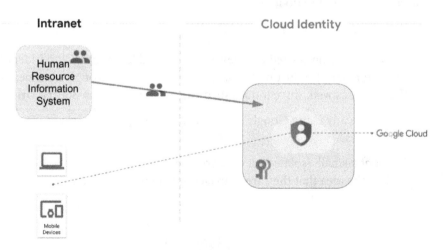

Figure 2.4 – HRIS manages users and Google handles IdP

This configuration allows the HRIS system to focus on what it's good at, HR, and lets Google handle something it's good at, *high-grade user authentication*. The biggest downside here would be that if your HRIS can also manage passwords, you aren't necessarily leveraging that. For some environments, that would imply that you have the same username in two systems, the HRIS and Cloud Identity, but the passwords wouldn't be the same. You'd log into local intranet systems through the HRIS and to Google Cloud through Google. It's either that or, more likely, you would make Google the only IdP, and your local systems would reach out to Google when logging into anything.

Here's a good link to check on the detail of setting this stuff up: `https://support.google.com/a/topic/7556794`.

This is when you should use it:

- You already have a solid HRIS in place and would like to preserve your HR system workflows as you move to Google Cloud.

- You're okay, and maybe even happy, to let Google handle the IdP, and have a plan to make Cloud Identity the only IdP for all in-and-out of cloud authentication – possibly because of the following point…

- There may be no existing centralized, on-premises IdP, and this gives you the chance to implement one.

This is what the login process will look like to a user (unchanged):

1. The user follows a link to a protected resource, such as Google Cloud Storage.
2. If their browser isn't already authenticated, then they are redirected to a standard Google sign-in screen where they enter their username and password. If 2SV is enabled, they are prompted to provide their key or code.
3. If the user authenticates, then IAM decides whether they are authorized to access the requested resource, and if so, they are then redirected back to the service they requested – in our example, Cloud Storage.

As I mentioned, one of the downsides with this approach is that some HRISs, as well as a bunch of other third-party apps such as AD, can manage both users and identities. What if we want to integrate one of those third-party services into Google Cloud? Well, there are several ways we can go about it.

Cloud Identity delegates all IdP and user management to an external (non-AD) provider

The phrase we're looking for here is: IDaaS provider. IDaaS providers range from specialized options, such as Okta or Ping, to some of the HRIS systems we discussed in the last section, such as SAP SuccessFactors or Workday, which can manage HR but also have components designed to handle IDaaS. Many of these **Single Sign-On** (**SSO**) services operate using an **Extensible Markup Language** (**XML**) standard called **Security Assertion Markup Language** (**SAML**), as shown in the following figure:

A third-party IDaaS provider acting as an IdP and managing users

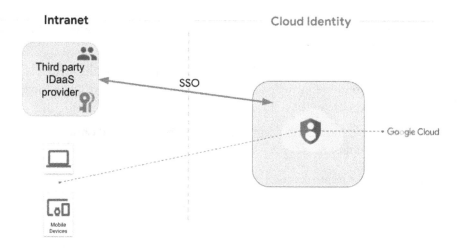

Figure 2.5 – A third-party IDaaS provider acting as an IdP and managing users

You know, I spent 4 years in the Marines, and I can tell you without any hesitation – the military has nothing on technology when it comes to acronyms.

For information on setting this option up, check out this link: `https://support.google.com/cloudidentity/topic/7558767`.

This is when you should use it:

- You already have an existing IDaaS provider with a set of users, and you don't want to reinvent the user management or authentication wheel.

- You don't need or want to synchronize passwords or other credentials with Google.

Behind the scenes, SAML works like this:

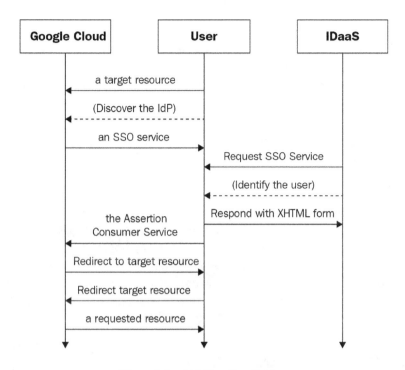

Figure 2.6 – SAML authentication

This is what the login process will look like to a user (SAML):

1. You follow a link to a protected resource, say Google Cloud Storage.
2. The browser gets redirected to the IDaaS provider.
3. If you aren't already authenticated, you log in at the SSO service, just like you do for everything else.
4. The provider passes a SAML response back through the browser to Google's **Assertion Consumer Service (ACS)**. The ACS makes sure that the response is appropriate, figures out which user you are, and then Google Cloud IAM decides whether you have access to the individual resource you requested. If you do, you get access to Cloud Storage.

This all sounds good, but what if that third-party service is Microsoft's AD? Well, that's a different kettle of fish.

Integrating Cloud Identity with Microsoft AD

I read a study once that said something like 95% of Fortune 500 companies use Microsoft AD as their primary IdP. So, chances are, if you're part of an established organization, you're probably doing your user identity management and your login through AD. How can Google Cloud integrate with an existing AD environment? There are four main ways, and I want to take time to explore each. Let's start with the two least commonly used options.

AD manages users and passwords on-premises, users are replicated to Google, and Cloud Identity then acts as the IdP for access to Google Cloud

This is probably the most simplistic form of AD integration. You let AD manage the users on-premises and use an automated tool from Google called **Google Cloud Directory Sync** (**GCDS**) to synchronize that user list with Google Cloud. Once Google Cloud Identity gets the user list, then users can log in and set up new passwords for use in Google Cloud. To be clear, this approach means that users have one username and password for everything on-premises, and then they will use the same username with a separate password to log into Google Cloud through Cloud Identity.

GCDS is software that you can download from Google: `https://tools.google.com/dlpage/dirsync/`. You provide a Windows or Linux server in your on-premises environment, install the software, connect it to Cloud Identity on one side using a new Cloud Identity superuser account (just used for GCDS), and connect it to AD using an account with minimal read-only permissions on the other. Then, you can use a **User Interface** (**UI**)-driven tool called Configuration Manager to set your configurations and let them run.

With GCDS, you can control the following:

- Synchronization frequency.
- Which part of your user domain gets synchronized, and within that, use rules to fine-tune the details.
- What user profile information will be passed to Cloud Identity when synchronization occurs.

All the data is passed to Google Cloud encrypted end to end. For more details, visit: `https://support.google.com/a/answer/106368`. Architecturally, this would look like the following figure:

AD manages users and Cloud Identity handles authentication

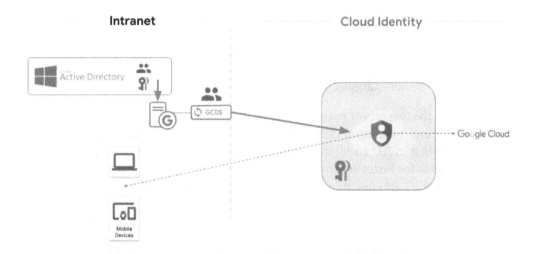

Figure 2.7 – AD manages users and Cloud Identity handles authentication

This is when you should use it:

- When you already have AD configured on-premises and the users who need to access Google Cloud are in it. So, AD is the one source of truth for which users exist, along with their group memberships, organizational structure, and so on.

- You are looking for the simplest AD to Cloud Identity integration, possibly because only a small subset of your users require Google Cloud access.

- You may want the Google access and IdP kept separate and external to your organization.

- You may be experimenting with integration and aren't ready to go all in just yet.

This is what the login process will look like to a user (as with any of the options using Google as the IdP):

1. The user follows a link to a protected resource, such as Google Cloud Storage.

2. Google recognizes the user because GCDS populated the user list. The user is redirected to a standard Google sign-in screen where they enter their standard username and Google Cloud-specific password. If 2SV is enabled, they are prompted to provide their key or code.

3. If the user authenticates, then IAM decides whether they are authorized to access the requested resource, and if so, they are then redirected back to the service they requested – in our example, Cloud Storage.

The issue here is the two sets of passwords. Even if a user manually sets them both to the same value, they aren't managed in a single place. If you need to update your password, you'd have to do that in AD and then again in Google Cloud Identity. In some cases, this approach can allow for better separation between your on-premises environment and Google Cloud, but it's also one more password to manage for your users.

If passwords being the same is your only worry, then you could synchronize them too.

AD manages users and passwords on-premises, users and passwords are replicated to Google, and Cloud Identity then manages access to Google Cloud

This is a slight variation on the last option. Here, you are still using GCDS to sync the users, but you're adding a separate application from Google to your on-premises environment, **Password Sync**: `https://support.google.com/cloudidentity/answer/2611842`. The nice thing about Password Sync is that it auto-syncs the passwords the moment the user or admin changes them in AD. The concern with Password Sync is that it's pushing all the passwords to Google. I know it's encrypted, Google-secured, and all that good stuff, but just the idea makes me a little nervous.

It looks like this:

AD manages users and Cloud Identity handles authentication - part two

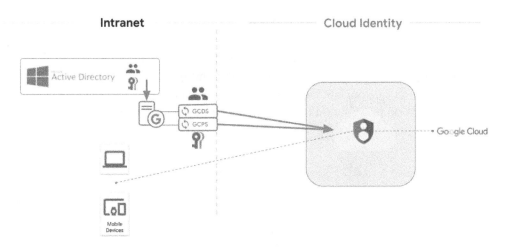

Figure 2.8 – AD manages users and Cloud Identity handles authentication – part two

This is when you should use it:

- When you already have AD configured on-premises and all the users who need to access Google Cloud are in it.

- You are looking for a basic AD to Cloud Identity integration, possibly because only a small subset of your users need Google access.

- You want the Google IdP part kept separate and external to your organization.

- You may be experimenting with integration and aren't ready to go all in just yet.

This is what the login process will look like to a user:

1. The user follows a link to a protected resource, such as Google Cloud Storage.

2. Google recognizes the user because GCDS populated the user list, and this time, Password Sync provided the password. The user is redirected to a standard Google sign-in screen where they enter their username and password.

3. If the user authenticates, then IAM decides whether they are authorized to access the requested resource, and if so, they are then redirected back to the service they requested – in our example, Cloud Storage.

Now that we have those two options out of the way, let's talk about the two most commonly used ways to integrate AD with Google Cloud.

AD manages users and passwords on-premises, users are replicated to Google, and Cloud Identity uses on-premises AD FS SSO for authentication

That's a mouthful, but it's easier than it sounds. This option starts like the last two – you download, set up, and configure GCDS. It focuses on what it's good at – replicating user information out of AD and into Google Cloud Identity. The big addition to the on-premises picture is **Active Directory Federation Services (AD FS)**, which is a Microsoft SSO solution. So, Google knows you're a user because GCDS replicated your user information into Cloud Identity, but the authentication is redirected back on-premises, where it's handled by AD FS. It looks something like this:

AD manages users and AD FS works as the IdP

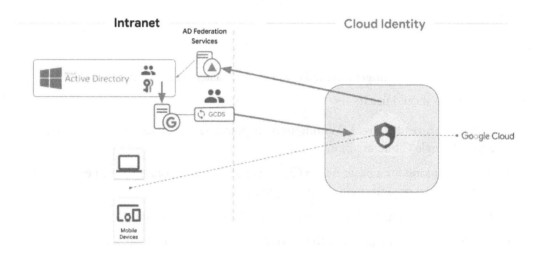

Figure 2.9 – AD manages users and AD FS works as the IdP

This is when you should use it:

- When you already have AD configured on-premises and the users who need to access Google Cloud are in it.

- You already have or are willing to set up AD FS to facilitate an on-premises SSO solution.

- You are looking for an AD-based solution where there aren't separate passwords for the different environments, and you don't want the passwords to be synchronized.

- You want to integrate Google into your existing AD FS SSO environment.

This is what the login process will look like to a user:

1. You follow a link to a protected resource, such as Google Cloud Storage.

2. If you are already authenticated into your on-premises AD environment, then SSO authenticates you behind the scenes. If not, then Google will redirect you to the AD FS authentication portal, as configured in your on-premises environment.

3. Once you have authenticated, Google Cloud IAM will determine whether you can access the protected resource. If so, you are then automatically redirected to the originally requested resource – in our example, Cloud Storage.

As you might imagine, the devil tends to be in the details. For in-depth coverage of federating with AD, as we've seen in this section, take a look at `https://cloud.google.com/architecture/identity/federating-gcp-with-active-directory-introduction`.

Unfortunately, any of these options which depend on GCDS, may run into issues when migrating some of the accounts to Google.

GCDS challenges

Have you ever signed up for a Google-owned service using your work email address as your username? Perhaps before the organization decided to use Google Cloud, you created a trial account, or maybe you signed up for Gmail using your work email as a secondary address? I'll lay odds that if you haven't, someone else in your organization has, and to make matters worse, they may not even work for you anymore.

When it comes to Google, there are two types of accounts – *managed* and *consumer*. Managed accounts are fully managed by an organization, through Cloud Identity or Google Workplace, such as my *gcp.how*. Consumer accounts are owned and managed by the individuals who created them, and they are used to access general Google services such as YouTube or Google Cloud.

There are several possible issues here, depending on the exact circumstances. For a full discussion, see `https://cloud.google.com/architecture/identity/assessing-existing-user-accounts`. I'm going to look at just a couple.

In our first pain-point example, Bob signs up for an account in Google Cloud and uses his bob@gcp.how corporate email address as his username. *gcp.how* has decided to move forward and create an IT presence in Google Cloud, with Cloud Identity acting as the IdP. You're in the process of setting up GCDS and you're working on the users you're going to migrate. Bob has an account in AD tied to bob@gcp.how, but Google already knows Bob as bob@gcp.how through the consumer account he created when he signed up as an individual GCP user.

The solution here isn't too bad. First, make sure you add and validate any variations of your domain name to Cloud Identity so that it knows about them. So, if *gcp.how* is your main domain but your organization also sometimes uses *gcp.help*, then add them both to Cloud Identity, with *.how* being the primary. Next, Cloud Identity has a transfer tool you can access (`https://admin.google.com/ac/unmanaged`), which allows you to find employee accounts that already exist as consumer accounts. You might want to download the conflicts and reach out to them yourself before initiating the transfer so that they can watch for the transfer request email and know what it is (not spam!) before it arrives. When ready, the Google transfer tool can send a transfer request to the user. When they accept, their account then moves from consumer to managed, and it falls under the control of the organization. For more details, visit `https://cloud.google.com/architecture/identity/migrating-consumer-accounts`.

Another variation on this same theme relates to former employees. Before *gcp.how* decided to move to Google, Malo used to work for them, but he left after some unpleasantness. Before he left, he was doing some experimentation with a private Google Cloud account tied to his email address, `malo@gcp.how`.

Now, the Cloud Identity transfer tool doesn't locate conflicting user accounts because they are in AD; it locates them because they are tied to a particular domain. You see the `malo@` account and, after a little research, you realize that he no longer works for *gcp.how*. A real concern here is that since he has an account in Google Cloud tied to a *gcp.how* email address, he might try a form of social engineering attack, perhaps by requesting access to a corporate project, and hey – the request is associated with a *gcp.how* account, right?

This is a pain point. In this case, you should evict the `malo@gcp.how` consumer account. The steps aren't bad and details can be found here: `https://cloud.google.com/architecture/identity/evicting-consumer-accounts`. Essentially, you create an account directly in Cloud Identity using the conflicting `malo@gcp.how` email account. You'll get a warning prompting you to ask for a transfer or to go ahead and create a new user. Create the new user with the same email address, and then immediately delete it. By purposefully creating a conflict and then deleting it, you will trigger an automated Google process that will force the existing `malo@gcp.how` account to change its credentials.

There are a few more pain points related to individuals using Gmail accounts for personal and corporate purposes, such as accessing documents and individuals who might use their work email address as an alternative email address on a Gmail account. These may be harder to troubleshoot, and details on handling these situations and more can be found here: `https://cloud.google.com/architecture/identity/assessing-existing-user-accounts`.

The final AD-related user and identity solution that I'd like to discuss is Azure AD.

Cloud Identity delegates all IdP and user services to Microsoft's Azure AD

There are a couple of reasons you might fall into this category. You may have moved some or all of your on-premises authentication into Azure as part of another cloud initiative, or you may still have on-premises AD, but at some point in the past, you set up federation from your on-premises AD instance to Azure AD. If this is the case, then Cloud Identity can pass the responsibility for all user management and IdP services to Azure AD:

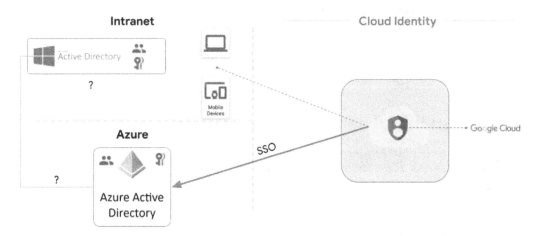

Figure 2.10 – Azure AD managing users and the IdP

For details on Azure AD federation, visit https://cloud.google.com/ architecture/identity/federating-gcp-with-azure-active-directory.

This is when you should use it:

- You already have, or are planning to implement, Azure AD as your authoritative SSO service.

- You want to provide a seamless SSO experience to your users across the on-premises, Azure, and GCP environments. Even AWS can sync with Azure AD.

- You would prefer your SSO service to be managed by Azure AD, rather than setting up and managing it yourself in AD FS.

This is what the login process will look like to a user:

1. You follow a link to a protected resource, such as Google Cloud Storage.

2. If you are already authenticated into your Azure AD environment, then SSO authenticates you behind the scenes. If not, then Google will redirect you to the Azure AD authentication portal and you will log in there.

3. Once you have been authenticated, Google Cloud IAM will determine whether you can access the protected resource. If so, you are then automatically redirected to the originally requested resource – in our example, Cloud Storage.

At this point, I'm going to assume that you've decided how you want to manage your users, what your IdP will be, and that you've created, or linked up at the very least, an initial set of key users. What constitutes a *key user*? Well, I'm about to tell you about a starter set of security groups that you need to create in Google Cloud. After I do, come back here and see whether you can find at least one user who would fit into each of those security groups.

Creating an initial set of security groups

Before we get to creating an initial set of groups and assigning users to them, let's have a quick side discussion on two security-related concepts – the **principle of least privilege** and **Role-Based Access Control** (**RBAC**). These aren't groundbreaking concepts to anyone familiar with security, but they are still worth mentioning.

The principle of least privilege is simply common sense – don't give users permissions that they don't need to do their jobs. If you work or have worked in an office building, then your swipe or key probably doesn't open every door in the building. Why? Because most jobs don't require a person to access more than a few specific building areas. If you're part of building management, security, or maintenance, then you might need unfettered access, but those are a handful of specialized positions, not average workers. Why not let everyone access everything? Because you're likely to get robbed if you do.

The same logic applies to Google Cloud. Remember in the previous chapter, when I talked about the huge list of services that Google Cloud offers? Well, you or someone in your organization is going to have to put in some time learning about how IAM security settings work in Google, and then research exactly what different roles in your organization need in terms of GCP access. We will discuss access control and IAM configurations in *Chapter 5, Controlling Access with IAM Roles*.

RBAC essentially says that when it comes to assigning security, think job roles instead of individuals. You might be your own special snowflake, but your job likely isn't. Most people are one out of a set of employees who perform a similar job function within the organization. So, instead of setting security for developer Bob directly, and then turning around and implementing the exact same security settings for developer Lee, put Bob, Lee, and the other developers on team six in the `dev-team-6` group, and set permissions for the group as a whole. It makes long-term management of security settings a whole lot easier. You've hired a new developer, Ada, into development team six? Just add her to the group, and bam – she gets the exact permissions she needs.

As far as the initial set of Google Cloud groups goes, Google recommends you start with six and that you eventually identify at least one initial user who you can slot into each. At this point, these groups won't have the Google Cloud IAM access settings to perform the jobs being proposed, but we will add that in a later chapter. Also, there's nothing that says you must use these exact groups, so feel free to tweak this initial list any way that you need. If you are looking for the wizard in Google Cloud to give you a nice green checkmark for completing this step, then you'll have to create at least the first three:

- `gcp-organization-admins`: Here will be your Google Cloud organization administrators, with full control over the logical structure of your organization in GCP. These are super-users, and there won't be many.

- `gcp-network-admins`: This will be your highest-level network administrators. These are the people who, among other things, can create VPC networks, subnets, control network sharing, configure firewalls, set routing rules, and build load balancers.

- `gcp-billing-admins`: Someone needs to be able to view and pay the bills, right? At least they do if you want to keep the Google Cloud environment lights on. There also needs to be someone who pays attention to your level of spending.

- `gcp-developers`: Ah, this is my old job. Here's where you'll put your top-level developers, who will be busy designing, coding, and testing apps in Google Cloud.

- `gcp-security-admins`: The core security-related people, with top-level access to set and control security and security-related policies across the whole organization.

- `gcp-devops`: Lastly, here you have your highest level of DevOps engineers, responsible for managing end-to-end continuous integration and delivery pipelines, especially those related to infrastructure provisioning.

Again, this list of initial security groups is in no way designed to be exhaustive or mandatory; it's just a nice place to start. If there are other high-level groups that would aid your organization in some way, you can add them now or at any time.

Now that you have a plan for your initial set of security groups, let's create them in Google Cloud. You can either use Google's new foundation wizard to create these groups (`https://console.cloud.google.com/cloud-setup/users-groups`) or you can manually create them yourself, perhaps adding a few groups that you've included in your plan. Here's how:

1. Log in to Google Cloud Console (`https://console.cloud.google.com/`) using your Cloud Identity super admin account created earlier in the chapter. If you followed my naming recommendation, then it's something like `gcp-superadmin@<your-domain>`.

2. Navigate to **Navigation menu | IAM & Admin | Groups**. The direct link is `https://console.cloud.google.com/iam-admin/groups`.

3. At the top of the page, click **Create**.

4. Provide a group name from your initial list; I'd recommend using the same name as the prefix for the group email address. That's why all the suggested group names are email address-friendly. If you want to, add an appropriate description.

5. Save.

6. As soon as you save the group, the **Group Details** page for the new group will appear and show you as the only current member. You should be logged in as the first organization admin and as the group creator; that account automatically becomes the owner of the group.

7. Use the **Add Members** button at the top of the page and add any users you have identified as members of this group.

8. Repeat *steps 1–7* until you have your initial set of groups created.

Very nice. At this point, you have your IdP and the user management figured out and set up, and you've created an initial set of high-level groups in Google Cloud, with a corresponding user or two in each. When ready, move on to *step 3* and set up admin access to your organization.

Step 3 – enabling administrator access

Up to this point, we've been working in Cloud Identity and Google Cloud using the emergency-only initial user, your equivalent to my Cloud Identity/Google Workspace super admin – `gcp-superadmin@gcp.how`. Remember how I said that this account should be reserved for emergency recovery and the like? It's now time to practice what I preached.

I'm a big fan of naming conventions. You come up with a convention that works, document it, and get everyone to use it. It helps with consistency and being able to identify what's what. If I were you, I'd create a Google or Word document somewhere that's accessible and sharable. Name it something like `Google Cloud Naming Conventions` and put a two-column table in it. Give the first column the heading `Resource`, and the second `Naming Convention`.

Next, add a row for the Cloud Identity super admins – `Cloud Identity super admin` has a nice ring. Next to it, come up with a naming convention such as `superadmin-first.last@<your domain>`. While you're there and naming things, add a second entry for your GCP organization administrators – how about `gcp-orgadmin-first.last@<yourdomain>`?

Resource	Naming Convention
Cloud Identity super admins	`superadmin-<first>.<last>@gcp.how`
GCP organization admins	`gcp-orgadmin-<first>.<last>@gcp.how`

Table 2.1 – GCP naming conventions

While we're moving through this book, I'm going to come back to our naming conventions document with a few other ideas about non-user-related resources.

Since a Cloud Identity super admin has more power than a GCP organization admin, I just followed my naming convention and created a pair of accounts for myself – `superadmin-patrick.haggerty@gcp.how` and `gcp-orgadmin-patrick.haggerty@gcp.how`. As, once again, these are high-security accounts, I enabled 2SV on both. As they are my personal accounts, instead of using a physical key, I configured my phone as a key. Modern phones, using Android and iOS (with the Google app installed), can act as keys (`https://support.google.com/accounts/answer/185839`). Be careful though – if you use the device as a key and the Google Authenticator app on the device as your backup, losing the device can lose both 2SVs at once. Make sure you get some recovery codes that you store securely. I use a password manager that can also store encrypted notes, and I store my recovery codes there. That way they are both encrypted and replicated across multiple devices.

Setting up the organization admin is something we'll do over in Google Cloud in a moment, but setting up the new super admin account is accomplished in Cloud Identity (`https://admin.google.com/`). In the left-hand menu, head down to **Account | Admin roles | Super Admin | Admins**. You should get a list displaying your current super admins with your emergency account listed. Click **Assign users**, type the prefix for the super admin user you created for yourself (`superadmin` if you're following my naming convention), and click **Assign Role**. Repeat the process if needed.

With the two new admin accounts created and with your new personal super admin up and running, log out from your emergency-use `gcp-superadmin` account and log back in under your personal super admin account (`superadmin-<first>.<last>@<domain>`).

Great! Your new Cloud Identity super admin is good to go, and you are ready to head to Google Cloud and get the corresponding organization administrator set up. Make sure that you are logged in as a Cloud Identity super admin and then head over to Google Cloud Console: `https://console.cloud.google.com/`. If you get a message about activating a trial account, dismiss it.

Now that we have our non-emergency accounts created and working, the *enabling administrator access* step of the Google Cloud foundational plan has three major things that you need to accomplish:

1. Verify that your organization was created correctly in Google Cloud.

2. Add your new organization admin account to the `gcp-organization-admins@<your-domain>` group that you created back in the last section and configure the permissions on the group correctly.

3. Grant other permissions that will be used in future foundation-laying steps.

Let's move through the list.

Verifying initial Google Cloud organization creation

Finishing the creation of your organization in Google Cloud is easy:

1. Verify that you are logged into Google Cloud Console (`https://console.cloud.google.com/`) and that you are using your new personal super admin account. If you are looking at Google Cloud Console, you can mouse over the circular avatar picture in the upper-right corner at any time to see which account you are currently using.

2. Navigate to **Navigation menu | IAM & Admin | Identity & Organization** and follow the instructions you find there.

> **Note**
>
> If you have just created this organization in Cloud Identity or Google Workspace, it may take a few minutes before it's picked up by Google Cloud.

3. In the project selector at the top of the page, verify that you can locate and select your organization, as shown in the following screenshot:

Figure 2.11 – Organization selected

With the creation of the organization verified, let's set up our top-level organization administrator group.

Configuring organization administrator group access

In an earlier part of this chapter, we created a collection of starter groups, including one for our organization admins named gcp-organization-admins. Currently though, the group has no permissions associated with it. We will change that now.

Google recommends that your GCP organization administrators be granted a group of security roles. As we mentioned previously, a Google Cloud IAM role is essentially a set of individual permissions needed for a particular job, as it relates to a particular part of Google Cloud. Google recommends that organizational administrators are assigned the following IAM roles:

- Resource Manager - Organization Administrator

- Resource Manager - Folder Admin

- Resource Manager - Project Creator

- Billing - Billing Account User

- Roles - Organization Role Administrator

- Organization Policy - Organization Policy Administrator

- Security Center - Security Center Admin

- Support - Support Account Administrator

Make sure you have Google Cloud Console open, are logged in using your super admin account, and have your organization selected, as shown in *Figure 2.11*. Then, follow these steps:

1. Navigate to **Navigation menu | IAM & Admin | IAM | Add**.

2. Verify that the resulting dialog states that you are adding a permission to your organization and not to an individual project. Mine currently reads `Add principals to 'gcp.how'`.

3. When you created your groups, each one had an email address, and if you were following my advice, then that address takes the form of `<group-name>@<domain-name>`, so for my organizational administrator's group (`gcp-organization-admins`), the email address would be `gcp-organization-admins@gcp.how`. Enter the email address for your organization admin group into the **New principals** textbox.

4. Use **Select a role** to grant the **Resource Manager | Organization Administrator** role. You can scroll to the **Resource Manager** category and then select the **Organization Administrator** role, or you can just enter `Organization Administrator` in the search box. However, if you use the search box, be very careful when selecting the role, as many are named very similarly.

5. Click **Add Another Role** and grant the **Resource Manager | Folder Admin** role.

6. Using the same process, also grant these roles:

    A. **Resource Manager | Project Creator**

    B. **Billing | Billing Account User**

    C. **Roles | Organization Role Administrator**

    D. **Organization Policy | Organization Policy Administrator**

    E. **Security Center | Security Center Admin**

    F. **Support | Support Account Administrator**

7. Verify that your permissions assignment dialog resembles the following figure:

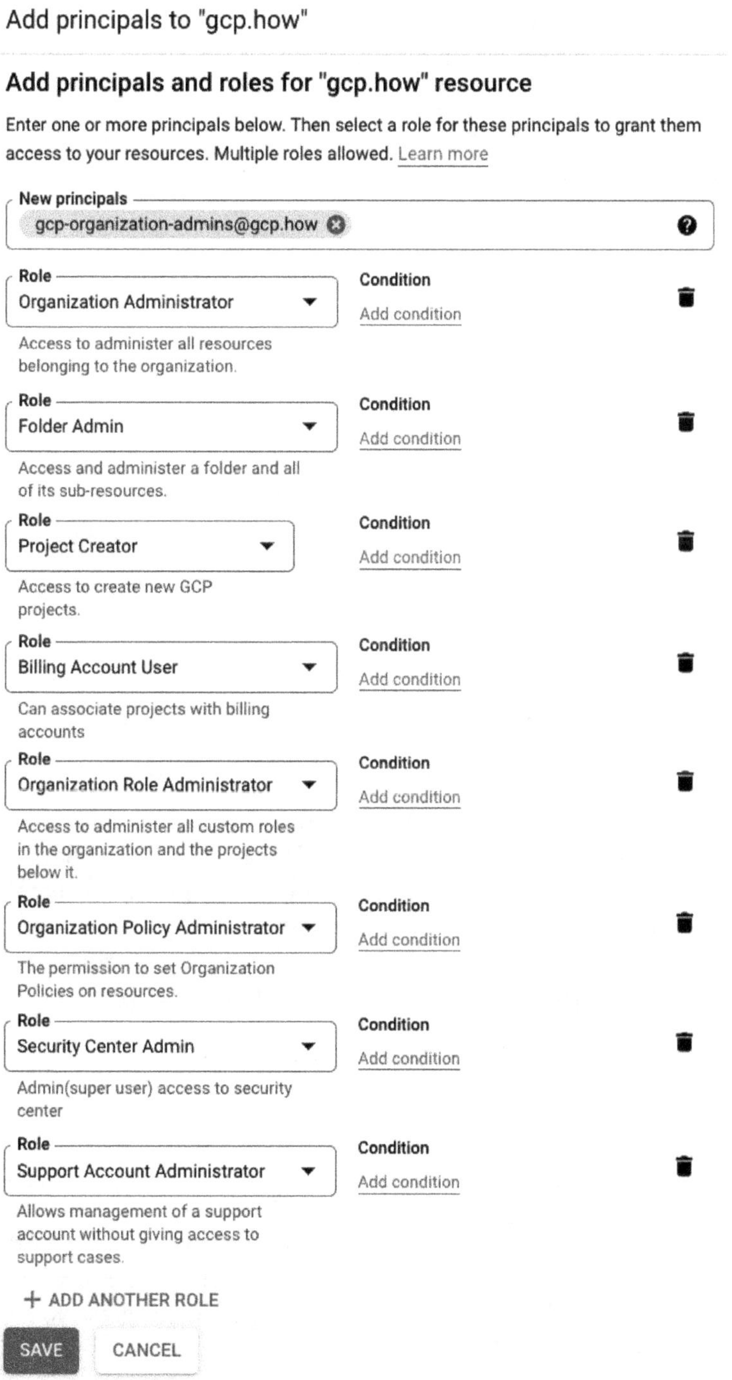

Add principals to "gcp.how"

Add principals and roles for "gcp.how" resource

Enter one or more principals below. Then select a role for these principals to grant them access to your resources. Multiple roles allowed. Learn more

New principals
gcp-organization-admins@gcp.how ✕   ❓

Role
Organization Administrator   ▾
Access to administer all resources
belonging to the organization.

Condition
Add condition   🗑

Role
Folder Admin   ▾
Access and administer a folder and all
of its sub-resources.

Condition
Add condition   🗑

Role
Project Creator   ▾
Access to create new GCP
projects.

Condition
Add condition   🗑

Role
Billing Account User   ▾
Can associate projects with billing
accounts

Condition
Add condition   🗑

Role
Organization Role Administrator   ▾
Access to administer all custom roles
in the organization and the projects
below it.

Condition
Add condition   🗑

Role
Organization Policy Administrator ▾
The permission to set Organization
Policies on resources.

Condition
Add condition   🗑

Role
Security Center Admin   ▾
Admin(super user) access to security
center

Condition
Add condition   🗑

Role
Support Account Administrator   ▾
Allows management of a support
account without giving access to
support cases.

Condition
Add condition   🗑

+ ADD ANOTHER ROLE

[SAVE] [CANCEL]

Figure 2.12 – New role assignments

8. Save the new settings. Google Cloud Console will take you back to your main IAM page where again, you can see that the security roles have been assigned to the group:

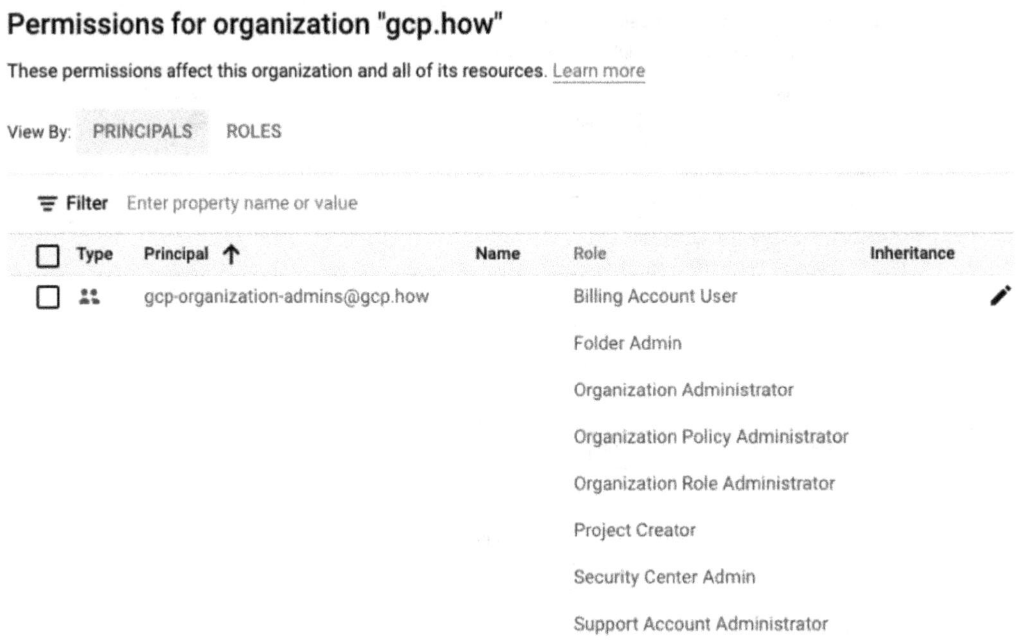

Permissions for organization "gcp.how"

These permissions affect this organization and all of its resources. Learn more

View By: PRINCIPALS   ROLES

☰ Filter	Enter property name or value			
☐ Type	Principal ↑	Name	Role	Inheritance
☐ ⠶	gcp-organization-admins@gcp.how		Billing Account User	✏
			Folder Admin	
			Organization Administrator	
			Organization Policy Administrator	
			Organization Role Administrator	
			Project Creator	
			Security Center Admin	
			Support Account Administrator	

Figure 2.13 – The newly assigned security roles

Now that our `gcp-organization-admins` group is properly configured, let's add our personal organization admin account as a member. Make sure that you have your personal organization admin account email address handy. If you're following my naming scheme, it should be named in the (where is that naming convention document?… ah, here it is) `gcp-orgadmin-<first>.<last>@<domain>` format. So, I'm `gcp-orgadmin-patrick.haggerty@gcp.how`. To add the account to the group, follow these steps:

1. Go to **Navigation menu | IAM & Admin | Groups | gcp-organization-admins | Add Members**.

2. Enter the new organization admin email address and click **Add**.

3. Repeat if there are any other organization admins that you'd like to assign.

Woo-hoo! Nice job. On Google's 10-step checklist, you can check the top 3 off. If you're using Google's wizard, then you'll see a way to mark each item as complete at the top of that step's page. You aren't done yet, but you're making good progress. Keep reading because there are more steps to do!

Summary

In this chapter, we started laying our foundation in Google Cloud by completing the first 3 steps in Google's 10-step recipe. Specifically, we set up our identity management in Cloud Identity, examined different ways to integrate external identity and user management systems such as AD with Google Cloud, created an initial set of security groups, and enabled administrator access to Google Cloud.

If you want to keep moving through the checklist steps with me, your personal tutor, keep on reading as we move on to *Chapter 3, Setting Up Billing and Cost Controls.*

3
Setting Up Billing and Cost Controls

If you're reading this book, then I'm going to go out on a limb and guess that you work with computers and IT. Do you ever get asked what you do for a living, and you run through a mental evaluation process wherein you decide how technical to be with your response? "*I work with computers*" is one of my typical go-to options. The person nods, but then comes the real question: "*I thought so – listen, I have this printer, and it won't print!*" You know about computers, so you must know everything in the world related to them, right?

Sigh. It always makes me feel sorry for medical doctors. How many times they must hear, "*Oh, you're a doctor? Listen – when I do this, it hurts. Why, Doc, why?*"

Why the story? Well, you might be tempted to think that billing is all about paying for stuff, so why not just skip this chapter? But like my computer printer story, even if you already work with Google Cloud, that doesn't always prepare you to configure billing properly or to implement good cost controls.

In this chapter, we are going to lay some foundational knowledge and then take the next major step toward laying our foundation in Google Cloud. To do so, we will be covering the following topics:

- Understanding billing terminology
- Step 4 – setting up billing and cost controls

Understanding billing terminology

To help us understand the various pieces and parts that play a role in billing, let's start with a diagram:

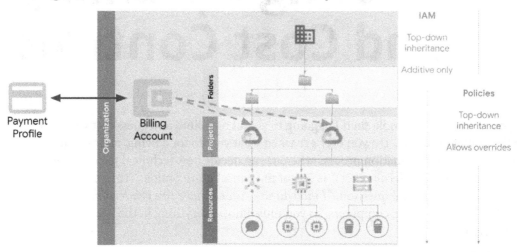

Figure 3.1 – Billing and resource management visualization

At the bottom of the preceding diagram, we can see what we want most out of Google Cloud – its resources. The word *resource* in English comes from the old French word *resorse*, which means to help or aid. That's pretty much what the bottom level does – it helps or aids our business by providing on-demand, scalable, metered services. From Pub/Sub messaging, to Compute Engine VMs, to Cloud buckets, the offerings are vast and powerful. Each of these Google Cloud services is logically managed by a single project.

Projects are special types of Google Cloud resources that let you directly enable, create, and logically group services. Every create command for a service, whether it's used via the UI, the command line, or a Google API, must be associated with a project. The project, in turn, owns and, through its associated billing account, pays for the resource. Projects and folders play a key role in organizing our resources, which we will discuss in the next chapter.

Projects, by necessity, are tied to a **Cloud Billing account**, which pays for whatever costs the services have incurred using one of the **Payments Profiles** (think a credit card or invoice) that's been configured. Payments Profiles are initially created as part of the Google onboarding process. Payments Profiles are Google-level resources that are managed outside of Google Cloud. They have users, permissions, and payment methods, all of which are managed at the **Google payments centre**: `https://pay.google.com`.

Great – with the general terminology under our belts, let's set up billing!

Step 4 – setting up billing and cost controls

As we've already mentioned, Google Cloud offers virtually any organization of any size a huge list of easily usable measured services. By measured, I mean that Google keeps track of the amount you use that service and then bills you accordingly. How such billing works is what we will discuss here.

It starts with how you pay

Google offers two main ways you can pay for services: self-service with a credit card, or via a monthly invoice that's paid by check or wire transfer. Organizations that want to move toward invoicing will need to configure a self-serve **Payments Profile** initially and then apply for invoicing once the organization meets the following criteria:

- The business must be at least 1 year old.
- Their Google spending must have exceeded $2,500 monthly for the last 3 months running.
- The company's name and billing address in the self-serve Payments Profile must match that in the company's legal registration.

The person applying for the invoicing must do the following:

- Accept the Google Terms of Service (`https://cloud.google.com/terms/`). This means they must have created, or at least viewed, a project in Google Cloud.
- Be a billing administrator for the organization.

Applications can be made at `https://support.google.com/cloud/contact/apply_for_invoiced_billing`.

Payments Profile user configuration options

You likely configured a self-serve Payments Profile indirectly when you first signed up for Google Cloud. As I mentioned earlier, the Payments Profile itself is not truly part of Google Cloud and as such, it has permissions and roles that are not related to Google Cloud IAM configurations. Two key job roles related to the Payments Profile that you need to identify are **Payments Profile Admin** and **Payments Profile Read-only Access**.

A Payments Profile Admin can view and manage payment methods, view payment accounts and invoices, modify account settings, see all the Google services associated with the Payments Profile, and make payments. Payments Profile Admins should be identified from the finance or accounting teams.

If it was you who initially set up your Google Cloud organization (organizational administrator), then you are likely already a Payments Profile Admin. If not, then you need to reach out to finance or accounting and determine who did the original setup.

To see whether you are already a Payments Profile Admin, go to the Google payments centre at `https://pay.google.com/`. If you get a page asking you to add a payment method, or if, when you look at the **Settings** page, you see **Account type** set to **Individual**, then you either aren't a Payments Profile Admin or you aren't logged in with the user you used when you initially set Google Cloud up. Mouse over the user's avatar in the top-right corner to see which account you are logged in under. Here's an example:

Figure 3.2 – My patrick@gcp.how user

If you believe that you are the Payments Profile Admin, then you may need to log in using the organizational administrator account you used when you first set up your organization in Google Cloud. For me, that was my `gcp-superadmin@gcp.how` account. If you are the Payments Profile Admin, the **Settings** tab should look similar to the following. Note that **Account type** is **Organization** (**Business** will appear for some) and that in the **Organization name and address** section, I can see my organization's name, `gcp.how`:

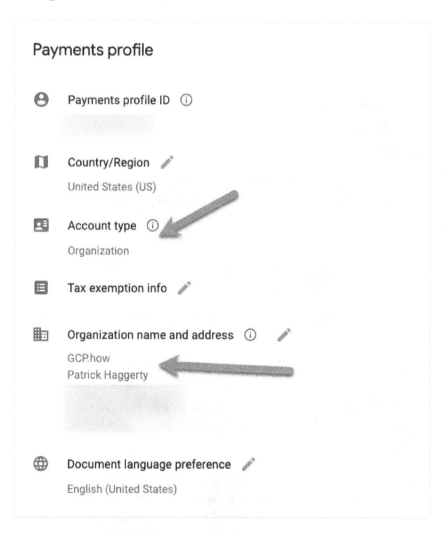

Figure 3.3 – GCP.how Payments Profile Admin

Below the general organization-related information and settings, you will see the current Payments Profile users list. If you have just set up your Google Cloud organization, then it's likely that you, as the organizational administrator, will be the only user. Below the current users, you can click **Manage payment users**, which will take you to a general Payments Profile user management area. This is where you can create Payments Profile users and set their permissions. Make sure you have at least two users with full (Payments Profile Admin) sets of permissions in case one is out of reach or no longer with the business. Typically, these users should be part of the finance or accounting teams. There are other individual permissions you can assign to personnel as needed. You will likely need to talk to your accounting/finance team to determine which users will need which, if any, of the following permissions:

- **Read access**: Read-only access to the Payments Profile

- **Edit payments profile**: Read and write access to the Payments Profile

- **Sign-up and purchase**: Pay for Google products

- **Manage users**: Add and remove users to/from the Payments Profile

- **Admin with all permissions**: All current and future permissions

> Note
> Payments Profiles span all the Google services that have been paid for by that profile, and they are not restricted to a single organization.

Regardless of permissions, Payments Profile users can also be configured to receive related emails:

- **All payments email**: All emails related to Payments Profile administration, receipts, invoices, statements, and other transaction-related messages.

- **Administrative payments email only**: Administrative emails, including suspensions, tax forms, Terms of Service updates, and account closures.

- **No emails**.

One – and only one – user will need to be configured as the **primary contact**. This is the person who Google will contact regarding payments-related inquiries. If you are the current primary contact, probably because you set up the organization in Google originally, then you need to consider if that makes sense. I'm on the tech side of the business, so even though I set up my organization in Google Cloud, I would likely not be the right person to contact regarding payment-related issues. I'd want to delegate that to someone in my finance department. Also, make sure the emails don't end up in spam!

Payments Profile configuration best practices

While Payments Profiles aren't one of the more complex parts of laying your Google Cloud foundation, there are still some things that you should keep in mind. To help, Google has a list of recommended best practices, including the following:

- Only use a single Payments Profile unless you need to separate personal from business profiles, profiles for multiple businesses you manage, or unless you need different profiles for different countries. New Payments Profiles may be created from the **Settings** page by clicking the pencil icon next to **Country/Region**.

- Set up at least two Payments Profile Admins but keep the number small overall.

- Make sure multiple individuals receive Payments Profile-related emails. Missed emails could lead to missed notifications, missed payments, and possibly even to account suspension.

- If you are using invoiced billing, configure multiple invoice delivery addresses for both paper and email delivery.

- For organizations not using invoiced billing, set up multiple payment methods on the **Payment methods** tab.

- Set policies and procedures to regularly review the payments profile. Is the list of *paid services* correct? Is the *primary contact* appropriate? Are the people getting *billing-related communications* correct? Is the *physical address* correct?

- Regularly review cost and payment history in Google Cloud (more soon), and the invoice if appropriate.

- If you have a dedicated finance team and a high level of Google Cloud spend, evaluate whether applying for invoiced billing makes sense.

With the Payments Profile configured, let's go back to Google Cloud and set up our Cloud Billing account.

Next comes Cloud Billing

As we mentioned earlier, projects help organizations create logical units of Google Cloud services. They encapsulate given sets of configurations, permissions, metadata, data, and other elements that comprise your cloud presence. Associated with each project is a **Cloud Billing** account, which will pay for any chargeable services that are used by the project through a given Payments Profile.

If each project needs a Cloud Billing account to measure and pay for spending, an early decision you need to make is how many Cloud Billing accounts your organization requires. Like Payments Profiles, less is usually more. Google recommends creating a single Cloud Billing account that you use to pay for all your Google Cloud spend. Besides making billing harder to track and manage, multiple billing accounts may not work the way you expect with **committed use discounts**. Committed use discounts are discounted prices associated with contractually agreeing to use resources for a specified time or amount (`https://cloud.google.com/docs/cuds`). The only exceptions to the single billing account recommendation would be if your organization needs to pay in multiple currencies or physically split charges for legal, accounting, or regulatory compliance reasons. Do not use multiple billing accounts because you think it will make spend tracking easier. Things such as resource labeling, billing reports, and BigQuery exports work much better when it comes to visibility.

At the very least, to set up billing for Google Cloud, you need to identify one organizational administrator and one billing administrator. If you are moving through the Google Cloud foundational steps in order and following Google's recommendations, then please recall the *Step 3 – Enabling administrator access* section of *Chapter 2, IAM, Users, Groups, and Admin Access*, where you created the `gcp-organization-admins` and `gcp-billing-admins` security groups. If you haven't already done so, make sure your two users have been added to their respective groups.

Billing-related IAM roles

Later in the book, we are going to dig into some details related to security roles in GCP, but since we are talking about billing, it's probably worth examining a few key related roles and how they may map to your organization.

Billing Account Manager

The **Billing Account Manager** has full control and access to all aspects of billing and billing accounts, including managing payments, viewing invoices, setting budget alerts, viewing spending, delegating billing-related roles, and communicating spending to parts of the organization that need it. This role is assigned at the Billing Account level and cannot be limited to a project.

The following users in the organization might be assigned this role:

- In small to medium-sized organizations, this could be some combination of CEO, CTO, and CFO, with the CEO or CFO likely managing and delegating billing-related tasks and managing Payments Profiles, while the CFO and/or CTO set budgets and view spending.

- In larger organizations, this might be someone in procurement or central IT. Besides managing Payments Profiles, they will likely set budgets and communicate spending to individual development teams.

Besides the Billing Account Manager, Google also offers a role with full read only access to billing data, the Billing Account Viewer.

Billing Account Viewer

The **Billing Account Viewer** is another role associated with the billing account, and it does exactly what you think. It allows users to view all cost and transactional data related to billing, including viewing invoices, billing reports, and all spending. This role may not be used at all in some organizations.

If used, the following users in the organization might be assigned this role:

- In small to medium-sized organizations, this might be someone in accounts payable that's responsible for approving invoices.

- In larger organizations, it could be someone approving invoices, or perhaps someone in financial planning working to interpret spending for others in the organization or C-Suite.

Between the Billing Account Manager and Viewer, Google offers a role to help view and manages costs, the Billing Account Costs Manager.

Billing Account Costs Manager

The **Billing Account Costs Manager**, like a viewer, can view all cost and transactional data related to billing, including viewing invoices, billing reports, and all spending. In addition, they have full control over budgets.

If used, the following users in the organization might be assigned this role:

- In small organizations, this role might not be used at all.

- In larger organizations, this should be someone in finance, accounting, or technical management who needs to monitor and make recommendations related to spending, as well as someone who needs to create, monitor, and manage billing budgets and alerts.

Another billing related role is the Billing Account User.

Billing Account User

The **Billing Account User** is also associated with the billing account itself and primarily lets someone associate a billing account with a project. Typically, this role will be granted to whoever is creating new projects for the organization.

The following users might be assigned this role:

- Small to medium-sized organizations may not use this role at all or may grant it to development team leads to help them create projects.

- In larger organizations, this might be a project lead or, again, someone in development.

Configuring the gcp-billing-admins group

In *Step 3 – Enabling administrator access* section of *Chapter 2, IAM, Users, Groups, and Admin Access*, we created the gcp-billing-admins group and identified at least one person for membership, but we didn't assign any permissions for the group in Google Cloud. In the official steps from Google, they set the permissions for this group in *Step 4*. The new setup wizard from Google, however, has recently started to assign group permissions in *Step 3*. Let's start by checking for proper group permissions before setting them when appropriate.

Assuming you are an organization administrator and are a member of the gcp-organization-admins group we created in *Step 3 – Enabling administrator access*, then you already have all the permissions you need to complete the following steps. However, since most Google Cloud organizational administrators aren't typically part of the finance and accounting units in their businesses, even if you are performing the following steps, make sure that you coordinate with your identified billing administrator so that they are aware of their new responsibilities. Let's get started:

1. Make sure you have logged into Google Cloud using your organizational administrator account. If you used my naming scheme from the previous chapter, then it should be in the following format: gcp-orgadmin-first. last@<yourdomain>.

2. Log in to the Google Cloud Console and navigate to the **IAM | Groups** page (`https://console.cloud.google.com/iam-admin/groups`). You may have to select your organization since groups are organizational resources and not project-specific. Check your group list and make sure you see the billing administrator group you created back in *Step 3 – Enabling administrator access* – that is, `gcp-billing-admins`. Take note of the group's email address (likely `gcp-billing-admins@<yourdomain>`) – you'll need that to verify group permissions in the next step.

3. With the group's existence verified, use the left-hand navigation menu to switch to the base **IAM** page. If permissions have already been assigned to the group, then its email address should appear as a principal on the page of organizational permissions, and the **Role** column should contain the **Billing Account Administrator**, **Billing Account Creator**, and **Organizational Viewer** roles, as shown in the following screenshot. If the group email address isn't in the list of permissioned principals or if the **Role** column doesn't display those roles, then you need to rectify that:

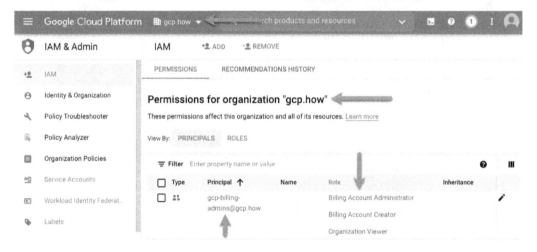

Figure 3.4 – The IAM page showing gcp-billing-admins and roles

If the group doesn't appear at all, click the **Add** button at the top of the page, enter the group email address in the **New principals** text box, and configure the group's security roles, as described next. If the group appears, but it doesn't have the security roles mentioned in the previous paragraph, then click the pencil icon next to the group to edit its permissions. Add/verify the requisite **Billing Account Administrator**, **Billing Account Creator**, and **Organizational Viewer** roles. Once you've done this, the group permissions should look as follows:

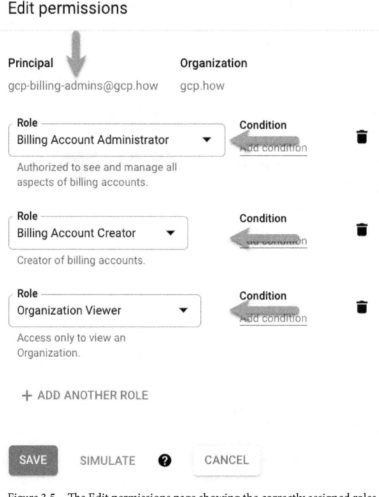

Figure 3.5 – The Edit permissions page showing the correctly assigned roles

4. Lastly, go to your Cloud Identity/Google Workspace admin page (`https://admin.google.com/`) so that you can assign your billing admin(s). From the admin home page, navigate to **Directory | Groups**. Mouse over your `gcp-billing-admins` group, as shown in the following screenshot, and click **Add members**. Add who you need to the billing admins group. Remember that you should always have at least two billing account administrators:

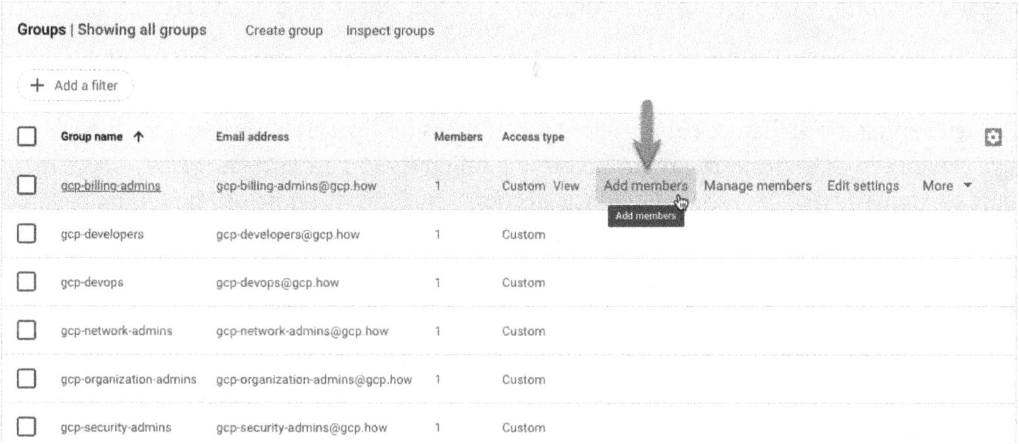
Groups | Showing all groups
Create group
Inspect groups
+ Add a filter
Group name
Email address
Members
Access type
gcp-billing-admins
gcp-billing-admins@gcp.how
1
Custom View
Add members
Manage members
Edit settings
More
Add members
gcp-developers
gcp-developers@gcp.how
1
Custom
gcp-devops
gcp-devops@gcp.how
1
Custom
gcp-network-admins
gcp-network-admins@gcp.how
1
Custom
gcp-organization-admins
gcp-organization-admins@gcp.how
1
Custom
gcp-security-admins
gcp-security-admins@gcp.how
1
Custom

Figure 3.6 – Adding members to the gcp-billing-admins group

Optional Recommendation

Previously, we discussed some of the other key billing-related Google Cloud IAM roles. It might not be a bad idea to start identifying people who may apply to those roles and set them up. You could, for example, create a group for Billing Account Viewers and add anyone relevant from the finance and accounting departments. Remember that this group would be able to see all the spending across all projects, so it would contain those that need to view organizational Google Cloud spending to aid with costing recommendations, investigating unusual spend, approving invoices, or communicating spending trends and relevant details with people who need to know. Once the group has been created and the members have been assigned, go back into Google Cloud IAM and assign the Billing Account Viewer role to the group.

Identifying your main billing account(s) and closing those you no longer need

Earlier in this chapter, I mentioned that Google recommends that you have a single billing account, bar a few exceptions. You, or whoever initially set things up for your organization in Google Cloud, may not have realized this best practice, and you may currently have multiple unneeded billing accounts. To check your billing account status, go to the main Google Cloud Billing page: `https://console.cloud.google.com/billing`.

First, identify the account(s) you want to keep and those you no longer need. For the account(s) you want to keep, decide whether the account names make sense, especially if you're keeping more than one. I'm following Google's best practice and only keeping a single account in `GCP.how`. My main account is called `My Billing Account`, which I don't particularly like. To rename a billing account, click on its name on the billing page to view its details. Using the left-hand menu, scroll to the bottom and select **Account management**. At the top of the page, click the pencil icon and enter a new name. Do this with any accounts you are keeping to clearly define what they are for.

Once you're happy with the names, it's time to clean out the billing accounts you no longer need. Before you can close a billing account, you will need to switch any projects it is paying for over to an account you are keeping. If the account that you want to remove is invoiced rather than self-served, then you'll have to contact Cloud Billing Support to close it: `https://cloud.google.com/billing/docs/support#contact-billing-support`.

Before moving projects to alternative billing accounts, another thing you will want to check for is commercial, non-free, products that have been installed from the Google Cloud Marketplace. You will need to make sure those items have been switched over or repurchased under the new billing account. For details, go to `https://cloud.google.com/marketplace/docs/manage-billing#manage_project_billing`.

To remove non-invoiced billing accounts, follow these steps:

1. On the main Google Cloud Billing page, `https://console.cloud.google.com/billing`, identify an account that needs to be removed and click on its **Overview** page.

2. Navigate to **Account Management** (where you can also rename accounts). One by one, click the triple dot menu next to each project to be removed from the billing account and **Change billing** to the main account. If you have a lot of projects to reassign, you may want to do some basic automation by using the `gcloud billing` command, which will link the projects to the main billing account (`https://cloud.google.com/sdk/gcloud/reference/billing`).

3. Verify that the billing account is no longer linked to any projects by visiting the **Account Management** page and ensuring that it is now empty.

4. Wait 2 days for any outstanding charges to hit the account. Charges may take up to 24 hours to appear on a bill, so the extra day will make sure they have all come in.

5. Settle the billing account by navigating to its **Overview** page. At the top, click **Payment Overview | Make a payment**.

6. Close the billing account from its **Overview** page by clicking **Close Billing Account** at the top of the page.

Pro Tip

On the main Billing page (`https://console.cloud.google.com/billing`), up at the top is a **My Projects** tab. Clicking this tab will provide you with a list of all your projects. Not only is this a quick way to see all the projects and which billing accounts they are associated with, but also at the right-hand side of the page is a down-arrow **Download CSV** button. This CSV file contains all the project names and their associated billing accounts. This may help you (automation) when you're trying to move a lot of projects to a central billing account.

With that, we have covered most of the core settings and changes that are required for a basic Google Cloud Billing foundation, but there are some best practices we need to discuss before moving on.

Google Cloud Billing best practices

My dad is one of those old-school southern gentlemen who does not discuss his finances – not with my mom, and certainly not with us kids. When I was growing up and things related to spending money happened, it was his deal, and he didn't want any of us to know the details. He's the same way today.

I'm sure some of you can relate, while others of you may roll your eyes and think, "*how odd.*" Would it surprise you that most organizations act much like my dad when it comes to spending in Google Cloud? The number one billing-related problem I see organizations make when it comes to GCP is not letting enough people in on the cost-making decisions.

In 2021, Google joined the FinOps Foundation. You can read about basic FinOps at `https://www.finops.org/introduction/what-is-finops/` if you like, but in a nutshell, FinOps advocates an organizational cloud spend culture where a central best practices group helps multi-skill FinOps teams decide and optimize cloud spending. That way, instead of someone in accounting looking at two products and making a decision based strictly on a Google Cloud line-item price, you get the costing knowledge out to a wider portion of the organization and involve people with multiple organizational skills in the decision-making process. It's not always the cheapest product that's best, and GCP products are rarely used in a completely isolated way, so involving more people with broader knowledge skill sets has a lot of benefits. Then, as you expand and grow in the cloud, you iterate the process and optimize as you go.

To put it bluntly, I want anyone making cloud-related product decisions to understand not just their technical capability set, but their cost ramifications. This means that if costs aren't transparent, then your employees won't know how to make cost-conscious decisions. I also want you to think bigger than just your Google Cloud bill. How much you spend on GCP services is important, but so are all the organizational costs related to cloud utilization.

Back in *Chapter 1*, *Getting to Know Google's Cloud*, I mentioned **Total Cost of Ownership (TCO)**. I defined TCO as what you pay Google, plus any related non-GCP costs. So, if you want a MySQL **database (DB)** in Google Cloud, you could spin up a VM in Compute Engine that's running Ubuntu and load MySQL yourself, or you could go to Cloud SQL and have Google spin up a managed MySQL instance for you. The Cloud SQL option would be a bit more expensive on your Google bill, but it would be substantially lower in terms of TCO. With Cloud SQL, Google manages the machine, the OS, the security, the database software, the backups, and the OS and DB patches for you. Compare that to what you are going to pay internally for all the same management on that VM you built, and Cloud SQL is likely to be significantly less expensive in terms of TCO.

To start controlling costs in Google Cloud, you need to know what you are spending.

Reading billing reports

Google provides multiple billing-related reports and information sets in the billing part of the Google Cloud Console. To view most of the reports, you will need to be a Billing Account Admin, Viewer, or Costs Manager. If you are a Project Owner, Editor, or Viewer, then you will be able to view a project-specific subset. Going back to the FinOps discussion, cost-related decisions should also percolate down to project managers and technical team leads – really, anyone that's responsible for making product choices in GCP.

To find the billing reports for your billing account, use the navigation menu and open **Billing**. If you have multiple billing accounts, you may need to click **Go to linked billing account**.

The **Overview** page will show your current spending, predicted end of month spending, cost trends for the last year, and the top spending projects and services, each with a link to a report with more details:

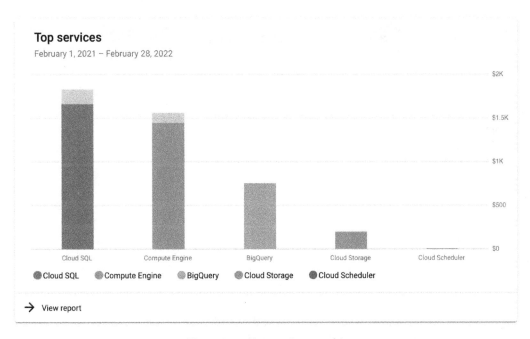

Figure 3.7 – Top services graphic

Reports is probably the most versatile of the GCP billing views (`https://cloud.google.com/billing/docs/how-to/reports`). Here, you can specify a filter for things such as date range, project, service, or SKU. Then, you can group the results by **Project**, **Service**, **SKU**, **Location**, or even by label keys. Labels are arbitrary key-value identifiers, and I'll discuss them later in this book. Here's an example where I'm looking at December 2021's spending for all my projects, grouped by service, and displayed as a daily cumulative stacked line chart. Here, I can easily see my three biggest spends by service:

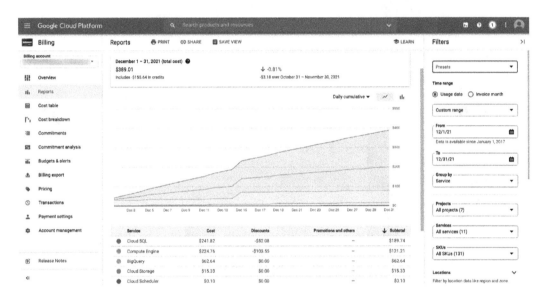

Figure 3.8 – Top spending services across all projects

Since the monthly statements are PDFs that provide limited details, you can use the **Cost table** view (`https://cloud.google.com/billing/docs/how-to/cost-table`), which gives you a customizable way to explore a detailed breakdown of your costs and credits for the selected invoice month. Details may also be downloaded as CSVs. Here's an example from December 2021. Notice how it has an itemized breakdown by project and service. As you check the sub-sections, you will see a running total in the popup summary box (in blue):

Figure 3.9 – Cost table for the Dev project's BigQuery spending

The **Cost breakdown** report (`https://cloud.google.com/billing/docs/how-to/cost-breakdown`) shows a filterable waterfall overview of your monthly costs and where you are currently realizing savings thanks to things such as committed use and sustained use discounts. Here, you can see my total would have been $544.65, but I saved some money thanks to committed and sustained use discounts:

Figure 3.10 – Cost breakdown showing current savings

The **Commitments** and **Commitment analysis** reports allow you to analyze your current committed use discounts and can recommend areas where further commitments could save you more. For details, see `https://cloud.google.com/billing/docs/how-to/cud-analysis`.

Once you have a good grip on what you are spending and how to get details from Google Cloud Billing reports, you should consider exporting your billing information to BigQuery for more detailed data science exploration.

Configuring daily billing data exports to BigQuery (sooner rather than later)

While billing console reports offer a lot of detailed information on how and where you are spending money in Google Cloud, its filters can only do so much. If you want to get your billing data someplace your data scientists can query with SQL or maybe even create custom BI dashboards for it, then your best option is to configure exports to BigQuery.

Tip – Set Up Billing Exports as Soon as You Configure Billing

Billing exports can only export daily billing information from the day you configure them. As a best practice, you should set up exports to BigQuery as a standard step when you're configuring Google Cloud Billing.

I'm going to walk you through how to manually configure billing exports now, but when we start infrastructure automation in the next chapter, you may want to update the project the exports are made into, since my automation will create a project expressly for that purpose.

BigQuery does two main things well: store data and query data. The queries are created in classic, ANSI 2011 standard SQL with a few extensions. But what about getting the data stored in BigQuery in the first place? That's what billing exports do for you.

There are three types of Cloud Billing data that you can export (`https://cloud.google.com/billing/docs/how-to/export-data-bigquery-tables`):

- **Standard usage cost data**: Account ID, invoice dates, services, SKUs, projects, labels, locations, costs, usage, credits, adjustments, and currencies.

- **Detailed usage cost data**: Everything from standard plus resource-level cost data related to Google Compute Engine VMs.

- **Pricing data**: Detailed pricing information.

For billing exports, you will need to specify the location that the billing data will be exported to. Google recommends using a new GCP project so that you can have a central location for anything billing administration-related. Within the project, you will also need to create a BigQuery dataset – that is, a location where tables of data can be stored.

To configure billing exports (`https://cloud.google.com/billing/docs/how-to/export-data-bigquery-setup`), follow these steps:

1. If needed, create a new project in Google Cloud to use for central billing administration support.

2. Within the project, go to the **BigQuery Data Transfer API** page by typing the API's name into the Google Cloud search box or by going to `https://console.cloud.google.com/apis/library/bigquerydatatransfer.googleapis.com`.

3. Make sure that the project selector has been set to your billing admin project. Then, **Enable** the API if needed.

4. Use the navigation menu (or the search box at the top of the Google Cloud Console) to navigate to the **BigQuery** page.

5. In the **Explorer** panel at the left of the BigQuery SQL workspace, click the triple dot menu next to your billing project's name and click **Create dataset**.

6. Enter a **Dataset ID** name; I'm going to use `all_billing_data`. I don't have any regulatory compliance to worry about as it relates to data location, so I'm going to store my dataset in `us`. I don't want my billing data to auto-delete, so I'm going to leave the **Enable table expiration** box unchecked and click **Create Dataset**.

7. Use the navigation menu to pull up **Billing**. If you have more than one billing account, make sure that you have selected the correct one. Then, click **Billing export**.

8. Click the **Edit settings** button for the type of billing export you'd like to configure. Select your billing admin project and the **all_billing_data** dataset you created. Then, click **Save**.

9. Optionally, enable any combination of billing exports by editing the settings for all the types you desire.

Again, remember that I will use automation to create a project for billing export storage in the next chapter, so if you want to wait on configuring exports for now, and come back next chapter, feel free.

Congratulations – you are now exporting your billing data! Take some time to explore the data table schemas at `https://cloud.google.com/billing/docs/how-to/export-data-bigquery-tables`. Google has some nice billing-related query examples at `https://cloud.google.com/billing/docs/how-to/bq-examples`.

Now that we have explored a couple of different ways to investigate spending, let's talk about what you are paying for.

Knowing what you are paying for

Traditionally, business IT costs tend to operate as a mix of **Capital Expenditures (CapEx)** and **Operational Expenditures (OpEx)**. CapEx deals with things such as buying servers, where you have a big outlay to purchase the hardware, but the monthly cost after that initial purchase is small, with the CapEx spending typically amortized over several years. On the other hand, OpEx covers things such as employee salaries, rent, taxes, and more. Moving to the cloud is a major change in the way businesses spend on infrastructure because most or all of the CapEx goes away, and everything becomes OpEx.

In the OpEx cloud model, you pay based on metered consumption. It might be by time, amount of data, operations executed, or various similar metrics. Understanding how Google charges you on a resource-by-resource basis is important.

As you are evaluating services in Google Cloud, look at the pricing pages for the products you are contemplating. If you go to `https://cloud.google.com/pricing/list`, you will see that Google has a link to the pricing page for each of its products. If you are working on deploying a new application and you know that you could use either Cloud Run or Google Kubernetes Engine, because either one could run the container-based application you are building, don't just learn the technical pros and cons, learn the differences in how you get charged both in terms of your Google bill and the related TCO. Now, we are making solid, well-informed FinOps decisions.

Something else to consider when you're selecting products is if they have a free tier, and if so, where it ends. Free tiers (`https://cloud.google.com/free/docs/gcp-free-tier#free-tier-usage-limits`) allow you to use certain GCP products up to some base level without charge. Free tiers may cover early development and testing and may even allow some low-use applications to continue to run in production for next to nothing.

Lastly, make sure that you investigate the committed use discounts I mentioned earlier. If you know that you will need to use Compute Engine, Cloud SQL, Cloud Run, Google Kubernetes Engine, or VMware Engine at some base level for long periods, then signing a contract for using those products could save you a lot of dough. For details, go to `https://cloud.google.com/docs/cuds`.

If you're attempting to predict pricing, fortunately, Google does have a price calculator that can help: `https://cloud.google.com/products/calculator`. This price calculator can give you very specific pricing information, but its results are only as good as the values you plug in. You're going to have to get into some of these products and roll around a bit to understand how much you will need and how the product generates spend.

That's another thing about FinOps, by the way – costing, as it relates to Google Cloud, needs to be an iterative process. You use a product and as you get better at it, you optimize your costs – and yes, sometimes, that optimization means you pick up and move to a different product.

For product-specific details related to controlling costs, go to `https://cloud.google.com/architecture/cost-efficiency-on-google-cloud`.

But you can't stop after simply knowing how products in Google Cloud spend – you need to set up some guardrails. Let's take a look at budgets and our first level of cost protection.

Creating and using budgets

A few years ago, a buddy of mine called me out of the blue and asked me if I'd like to spend a couple of weeks in Greece. It seemed he was going to Tinos, an island out in the Aegean Sea, to learn how to carve marble, and he invited me to keep him company. I'm about as artistic as a post, but a couple of spring weeks in Greece? *Hell yeah.*

While my friend was learning to carve beautiful Greek marble (and making his hands extremely sore), I rented a car and drove it all over various islands off the coast of Greece. Now, the Greeks aren't big on guard rails. I'd be cruising along some road on the side of a mountain where, at the edge of the road, there would be a cliff that dropped away to the rocks and blue waters of the Aegean, and there wouldn't be a guard rail in sight. To say it made me a little nervous would be an understatement. One time, I met this bus on a curve, and I was on the outside, as was the cliff…

It's true – your whole life really can pass before your eyes.

Learning things the hard way has its benefits, but I'm telling you this for your own good so that you don't make the same mistakes as other cloud newbies. People using the cloud need to know how their spending works regarding the products they are using, and they also need to know how to implement cost controls.

Google Cloud Billing budgets are alert notifications related to spending, and they are typically sent out through email to up to five recipients. They are Google Cloud's way of tapping you on the shoulder and saying, "*Hey, you've spent (or are on track to spend) x% of your specified budget amount.*" They can be attached to services, projects, and/ or entire billing accounts, and you can specify monthly, quarterly, yearly, or custom calculation periods.

Now that you know what a budget alert is, let's look at an example where using one could have saved an organization a lot of money.

Once upon a time, I did some work for an organization where a genetic researcher created a Cloud Storage bucket. Cloud Storage bills per byte for what you store, and you can select one of four storage classes, depending on your Cloud Storage usage pattern. The standard storage class is the most expensive for storage, but you pay no penalty for accessing files. However, the other three storage classes each bill for storage and access. For example, a regional storage bucket that's been created in the us-central1 region with a standard storage class currently bills $20 a TiB for storage and there's no access penalty for using data. If you store a 1 TiB file and read it 10 times in the month, you pay $20. The Archive storage class for the same bucket in the same region bills storage at only $1.20 a TiB, but there's a $50 per TiB access penalty. This storage class is for files that you hope you will never use or certainly don't need to use more than once a year. Store that same 1 TiB file and read it 10 times in a month and your bill is going to end up being $501.20, almost entirely thanks to the access penalty.

Can you guess where my story is going?

So, the geneticist is storing a lot of files in Cloud Storage – big, sequenced DNA files if you can believe it. The storage that's required for a single person's fully sequenced DNA is about 200 GB. So, if you have hundreds or thousands of people and you're running some ML, then that's lots of data. The geneticist is thinking about storing all this data, so she looks at the Cloud Storage price sheet (`https://cloud.google.com/storage/pricing#price-tables`) and she picks Archive because it's by far the least expensive. She doesn't read the whole article (boring), so she never notices the *Retrieval and early deletion* section, which discusses data retrieval costs. At the end of the month, she and her manager, as well as her manager's manager, all are a bit surprised by the $5,000 bill for storage.

Have you ever heard Einstein's famous quote, *"Everybody is a genius. But if you judge a fish by its ability to climb a tree, it will live its whole life believing that it is stupid."* I think it applies quite nicely here.

There are a lot of things wrong in this story. Why was the geneticist making decisions on Google Cloud architecture, without the aid of someone who knows Google Cloud as well as she knows genetics? There were a lot of researchers doing very similar sorts of things, so why wasn't automation in place to build out the infrastructure, perhaps in the form of a Google Cloud project per researcher? Why in the world were there no budgets in place to spot unusually high spending before the bill came at the end of the month?

As a side note, this is exactly the sort of example where FinOps makes sense. I need a geneticist on the FinOps team because they know what the business is trying to do a lot better than me. That 200 GiB per sequenced human metric came from her. But there also needs to be someone on the team who knows how spending in Cloud Storage works, as well as how spending works in all the other products related to Cloud Storage.

As I mentioned previously, budgets would have helped here too. Guesstimate the monthly spend for your whole presence in Google Cloud and set it as a budget on the billing account. Guesstimate your total spend for every project you have, and you can set that as a series of budgets too. When you are using a particularly heavy spending service, or are experimenting with a service you don't know, set a budget on that specific service. Budgets can be attached to specific amounts, forecasted amounts, and percentages of the same.

It works like this: *"I think I'm going to spend a total of $1,000 this month in Google Cloud."*

Great – then let's set a budget for that $1,000, with alerts that fire at 25%, 50%, 75%, 90%, and 100% of actual spend, and at 75%, 90%, and 100% of the forecasted spend. That might be a bit heavy on alerts, but especially when you're starting, it can be useful to see how close to your budget you're going to be. If your 100% forecasted and 25% actual alerts both fire in the first 3 days of the first week, then you're not going to come close to your budget.

Initially, you may need to adjust your budgets and alerts daily, moving to weekly, then moving to a few times a year. You should set a policy for regular budget reviews while checking the quality and applicability for each alert, modifying some, and weeding others out completely.

> **Warning – Budgets Trigger Alerts, They Don't Limit Spend**
>
> When you run through 100% of a budget, it will not stop or slow down your total spending. Budgets are alerts related to actual and forecasted spending; they don't turn off the tap. If you need to stop spending, consider creating a budget that sends out an alert message through Pub/Sub to a Cloud Function or Cloud Run container with a Pub/Sub trigger attached. This code could power down a specific service. One quick trick to stop spending is to remove the billing account from a project. This would cause all the products in the project to shut down until a new billing account is attached.

To create budgets and budget alerts, you need to have the **Billing Account Administrator** or **Billing Account Costs Manager** IAM security role.

To create a budget in the Google Cloud Console, from the navigation menu, go to **Billing**. From the left menu, navigate to **Budgets & alerts**. Then, at the top of the page, go to **Create Budget**. Provide the following information:

- **Name**: Make it descriptive so that you can recognize it in a list. My initial alert is going to be **Total Monthly Spend**.

- **Scope**: **Monthly** alerts for **All projects** and **All services** is a good place to start. Later, once you have the broad alerts set up, you can come back as needed with alerts for specific projects and even for specific services. If you are thinking about using a service that's new to you, that's probably an excellent time to add a service-specific alert constructed to your predicted spend numbers.

- **Amount**: Alerts tied to specific dollar amounts are the simplest. In the long run, adding alerts relating the current month's spend to last month's spending may also be helpful.

- **Actions**: Initially, alerts with a good spread of actual as well as forecasted percentages can be really helpful. Over time, you may want to trim out some of the notification thresholds.

Email alerts to billing admins and users is good, but you may wish to consider linking in monitoring email notifications as it will allow you to arbitrarily choose a notification technology (email, Slack, and so on) and the people who get notified through it.

> **Warning – Make Sure People are Seeing the Alerts**
>
> One major gotcha with billing alerts is tying them to individual email addresses. If you send all your alerts to Pat in finance, and Pat is on a cruise or has moved on to that job in IT she's always secretly wanted, then she may not see the alert, and alerts to bad email addresses aren't reported anywhere in GCP. Instead of alerting individuals, you may consider alerting a group of people. When users come and go from finance, adding and removing them from the group can ensure that someone appropriate is always notified.

On the right-hand side of the budget window, you'll see a **Cost trend** chart, as shown in the following screenshot. As you refine the budget's scope, the chart will update to allow you to see historic amounts based on your selections (projects, services). The chart works as a good sanity check, displaying a red dashed line where your configured alert would hit 100%:

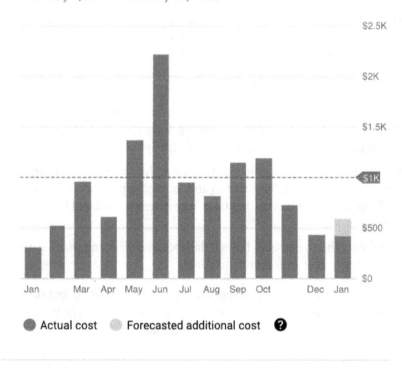

Figure 3.11 – Cost trend showing monthly spending with a $1k budget line

In addition to setting budgets to notify your billing admins and key personnel on what's happening with spending, you may also want to consider creating alerts in Monitoring.

> **Warning – Google Updates Billing Data About Once a Day**
>
> This means it's completely possible for you to spend lots of money before you get any alerts. You should know how products generate spending, and how the configurations you select relate to spending, along with product-specific spending recommendations, before enabling those products in your project.

Creating alerts in Monitoring

We will cover monitoring in detail in a later chapter. For now, **Google Cloud Monitoring** allows you to collect metrics about services running in Google Cloud. A metric might be the CPU utilization for a running VM. The data is aligned (bucketized) and you get back data points on some schedule, once a minute most commonly. Not only can you throw those data points on a nice visualization (dashboard), but you can also use them to create alerts.

We need to talk a lot more about monitoring and alerts, but I have mentioned it here in the billing section because while billing alerts take, on average, 24 hours to update, things in Cloud Monitoring tend to update in a matter of minutes.

Here's how monitoring could help you spot overspending:

1. You've identified a new Google Cloud service, *X*, and are starting to experiment with it. You know how it generates spending, have read a bit about keeping costs down, and you may have even created a service-specific budget alert just in case.

2. You take a peek at Google Cloud's metrics page (`https://cloud.google.com/monitoring/api/metrics`) and locate the product. You are looking to see whether there's a metric that takes measurements related to how service X generates spending.

3. If you find one, then you could build a dashboard and/or monitoring alert to notify you if that metric goes over some threshold.

Monitoring alerts can't always help you spot out-of-control spending, but sometimes, they can give you better visibility into things before the costs get to the billing alerts. Something else that may help you cap spending is quotas.

Setting up quotas

Resource quotas are artificial caps that limit how much of a given service you can use, in a given time, within a given project. Quotas are designed to limit unintentional resource overruns and their corresponding costs. There are two major types of quota:

- **Rate quotas**: Limits the number of times you can use a service API in a particular period.

- **Allocation quotas**: Limits the amount of something you can use, such as the total number of VMs or networks, within a given project.

For quota-related details, most products in Google Cloud have a **Quotas and limits** page you can easily find by searching for it. You can view and set quotas in a project by going to the navigation menu and selecting **IAM & Admin | Quotas**.

Here's how to make quotas work for you.

Imagine that you are about to start using BigQuery. You've done your due diligence and you know how BigQuery pricing works (`https://cloud.google.com/bigquery/pricing`). You've determined that you aren't going to use BigQuery enough to make flat-rate pricing worth the money, so you'll be paying per query at $5/TiB data processed. You also know how to tell how much a query costs using the UI, and you have your budget in place – perhaps even set at the BigQuery service level – but you'd still like to limit the total daily spend.

Here's where a quota helps.

You head over to Google Search, where you enter `GCP BigQuery quotas and limits`. In the results, the first choice is `https://cloud.google.com/bigquery/quotas`. You do a little reading, and you see there's a *Query usage per day* quota, and that, by default, *it has no limit*.

Yup – by default, there's no limit on how much you can spend on BigQuery queries in a single day. You've come up with a $1,000 query? It's legal. If you have a billing budget attached to the BigQuery service, with the budget amount set to $25, then sometime in the next 24 hours when Google updates the billing data, and that single $1,000 query appears on the bill, you'll get all your alerts at the same time. If you decide to investigate, then you'll realize that you overspent, just a bit. Granted, as far as I'm concerned, anyone with the power to use BigQuery should have been educated on how to tell how much a query is going to cost before it's run, but an actual limit might be nice here.

Time for a quota.

You navigate to the project where you are going to start using BigQuery and use the navigation menu to go to **IAM & Admin | Quotas**. In the quota filter box, search for `Query usage per day`. When you see the quota in the list, check it off, and, at the top of the page, click **Edit Quotas**:

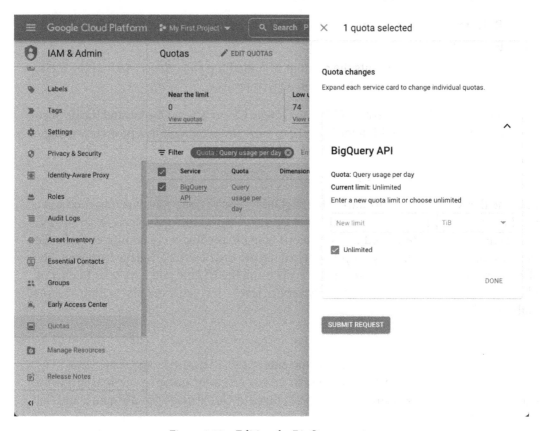

Figure 3.12 – Editing the BigQuery quota

If you get billed $5 per TiB, and this quota is looking at being queried data daily, then if you enter 5 and set the unit to **TiB**, that would cap your max BigQuery usage to $25 a day. When you exhaust that quota, the queries will start to throw an "*out of quota*" error message.

This isn't a bad trick, and it's one that far too few people take advantage of.

Great job! You should now have a solid Google Cloud Billing foundation in place in terms of budgets, quotas, and billing exports, and you should also have the skills to know what you're spending money on and why. You can safely tick off *Step 4* in our 10 steps to laying a foundation in Google Cloud. Great work – but don't stop now.

Summary

In this chapter, we continued laying our foundation in Google Cloud by completing *Step 4, Setting Up Billing and Cost Controls*, from Google's 10-step recipe. You now have a better understanding of how Google Cloud Billing works and can configure both Payments Profiles and billing accounts. You know how to find billing reports, can create budgets and budget alerts, and can cap usage by creating quotas. You also learned a little bit about the new FinOps approach to keeping cloud costs under control and have a path to breaking down the cost knowledge walls your organization likely has. This will help keep more of the organization in the loop when it comes to what things cost in Google Cloud. You've done a fantastic job and just think – you're almost halfway there!

If you want to keep moving through the checklist steps with me, your tutor, then please move on to *Chapter 4, Terraforming a Resource Hierarchy*.

4
Terraforming a Resource Hierarchy

Imagine that your business is based in a large city, and management has decided to move all the employees from several small satellite locations to a centrally located office building. Your boss calls you in and tells you that you are going to volunteer to head the committee, who gets to decide where everyone is going to sit in the new building. Aren't you lucky? So, how would you go about it?

There are a lot of right ways you could organize the employees, but if it were me, I'd say we need to attack the problem on two main fronts: organizational and security.

A good place to start would be to examine the natural employee order you already have in place, all while identifying major job types and existing teams and groupings. But looking at the employees strictly from an organizational perspective isn't going to be enough. This is mainly because, in real life, people don't all sit in locked offices with security at the office door and nowhere else. Some people are in private offices, some in cubes, some in open work areas, and wherever they may sit, they tend to work and interact in groups. To facilitate that, building security is typically handled in multiple layers, from the entrance to the floor, to the building section, and sometimes down to the room itself. Organizing the employees so that the groupings make sense from both an organizational and a security viewpoint will drastically simplify things.

Google Cloud may not be an office building, but there are a lot of conceptual similarities when it comes to setting up our resource hierarchy and, later, your access control.

We've mentioned a few times that Google Cloud projects are logical collections of services with a billing account attached, as well as that folders are logical groupings of projects and other folders that are used for organization purposes and to define trust boundaries.

I can't tell you how many times I've shown up at an organization and discovered that they had a hundred projects and not a single folder in sight. In my mind, I was thinking, *"You're doing it wrong."* Not that working without folders can't work – it's that using only projects means that all the security and policy settings are handled on a project-by-project basis. You can't tell me that there aren't groups of projects with the same organization policies and security settings, perhaps managed by the same teams? Instead of a hundred projects, each with their own settings, it would be much easier to group projects with similar policy configurations into folders, apply the policies at the folder level, and then let those policies inherit down to the projects themselves. In the long run, that's so much easier to manage.

Time to get to work! In this chapter, we're going to introduce infrastructure automation with Terraform, and then use Terraform to help set up our resources. We will be covering the following topics:

- Automating infrastructure with Terraform
- Step 5 – creating a resource hierarchy to control logical organization

Automating infrastructure with Terraform

When I was in high school, my dad got his pilot's license and bought a small, twin-engine Comanche. Imagine a smallish four-seater car with wings, with some luggage thrown behind the two back seats – that's pretty much what it's like. The plane rolled off the assembly line in 1965, but my dad upgrades it constantly. Commercial pilots would drool over that little plane's electronics. Honestly, I'm not sure if he likes flying as much as he likes upgrading his plane.

When I was a senior in high school, I remember that we took a trip. As the pilot, he was in the left front seat, and I was riding shotgun in the seat next to him. We were getting ready to take off and out he pulled out a laminated card of steps. It looked like original 1965 equipment.

I said, *"What's the matter, can't you remember how to take off?"*

He looked at me through these big black rectangular glasses and deadpan said, *"Sure, but this way, if some smartass interrupts me in the middle of things, I'm much less likely to forget a step and kill us all."*

We have all worked through something using a checklist. Checklists are both good and bad. They're good because they remind us what to do, and in which order, but they're bad because when we go through them manually, there's always a chance we will forget something. As the list gets longer, and as we go through it more times, the chance of us missing something increases. Going through checklists can also be quite tedious.

Google calls most cloud-related checklists **toil**. In Google's original book on building and running production systems, **Site Reliability Engineering** (SRE), which you can read for free at `https://sre.google/books`, Google defines toil as, *"the kind of work tied to running a production service that tends to be manual, repetitive, automatable, tactical, devoid of enduring value, and that scales linearly as a service grows."*

Planning a resource hierarchy of projects and folders is an art and a science. Building it, and managing that structure over time, is toil.

Imagine planning the resource hierarchy for a data processing application you are moving into Google Cloud. Throughout the week, various company websites generate a folder full of files containing website usage data from direct web visitors and various banner advertisement services. Every Saturday, the application runs a Spark job that extracts and transforms the data in the files and throws the results into a big on-premises database. During the week, data scientists run queries against the database to study website usage trends. Since the table size has increased (it now contains over a trillion records), the query efficiency has dropped. A decision has been made to move the whole kit-n-caboodle to Google Cloud. The plan is to collect and store the files in Cloud Storage, do the Saturday Spark job on an ephemeral Dataproc cluster, and store and query the results in BigQuery.

From a project perspective, how would you approach this problem? Build a single project and do everything there? You could, sure, but Google would recommend that you build at least three projects – one for development; one for non-production staging, testing, and QA work; and one for the production system itself. To make things toil-worthy, you don't just have to create three identical projects. At first, you don't know all the infrastructural pieces that you will need, which means elements could change over time. This means that not only do you need to build three identical sets of infrastructure, but you will need to evolve them as your design changes and make sure they are as similar to each other as possible.

Another related decision would be, are you going to store the data, process the data, and query the data out of a single set of projects, or are you going to store and process the data in one set, then move the results into a more centralized data warehousing project? So, you may not be talking about three projects – there could be considerably more.

Now, imagine the problem multiplying as the number of applications and systems expand.

Toil.

Let's see how Terraform and other **Infrastructure as Code (IaC)** technologies can help.

Infrastructure as Code to the rescue!

If you need to create and manage three identical projects, the manual way would be to use checklists, build your initial development project, and document all the infrastructure. As the application evolves through various infrastructural changes, you must keep tweaking the checklist (and please lord, don't miss documenting anything). Hold off on the non-production and production projects until you get close to your minimum viable product. As late in the game as possible, use your checklist to create the non-production project. Do your testing and QA work, and if any changes are needed, then document them, and retrofit them into development and non-production until you are ready to go to production. Finally, take your checklist and build the production project.

Matter of fact, where's the new guy? Following a super long and detailed checklist sounds boring, so why don't we get him to do it?

A much better approach would be to create a script that, in effect, is an executable version of a checklist. Basic scripting using bash files based on `gcloud` and other commands may not have much elegance, but they are tried and true and can get the job done. Like any scripting, if possible, store the files in a version control system such as Git. Even better, why not automate how the scripts are applied with a **continuous integration** (**CI**) and **continuous delivery** (**CD**) pipeline? Now, we're getting code-like – that is, creating our infrastructure with code.

And thus, **Infrastructure as Code (IaC)** was born. IaC is the process of provisioning and managing your cloud infrastructure using scripts and other configuration files, which are typically stored in Git version control and ideally applied using CI/CD pipelines.

Though the bash scripting approach will work and is quite common in a lot of environments, it lacks many key benefits that are brought to you by tools specifically designed for IaC. Bash scripts, while flexible, are also wordy and complex when it comes to creating infrastructure. It's even harder to code scripts so that they're flexible enough to adapt to inevitable change, without you having to tear out the whole environment and start from scratch. While there are several IaC tools, the two most commonly used with Google Cloud are Deployment Manager and Terraform.

Deployment Manager is an IaC tool built right into Google Cloud. It came out as **Generally Available (GA)** in 2015 and has been a core GCP tool ever since. Using a mix of YAML, Python, and jinja files, it can automate how infrastructure is created, managed, and removed. Though it is nicely integrated right into Google Cloud, it has always felt a bit clunky, and its popularity has never really taken off. Also, with more and more companies operating their infrastructure over a mix of on-premises and cloud environments, a tool with a broader possible application does have its draw. Enter Terraform:

Figure 4.1 – Terraform logo

Terraform (TF) is an open source IaC tool created by HashiCorp (`https://www.hashicorp.com`). HashiCorp provides several freemium products, including Terraform, Packer, and Vault. TF has more than 1,000 community-created providers, allowing it to create and manage infrastructure in most cloud and on-premises environments. TF configuration files are written in a human-readable format known as **HashiCorp Configuration Language (HCL)**, and the overall creation and application of those files can be accomplished using the same steps, everywhere.

While this information is accurate, it's also sort of misleading. Yes, TF can create and manage infrastructure in all the major cloud providers, as well as most on-premises environments. Yes, the major steps you go through to use TF in any environment are always the same. And yes, the majority of your work will be creating and using HCL files.

But.

The HCL to create a VM in AWS, and the HCL that's used to create a VM in GCP, while sharing a lot of logical similarities, are quite different. When I hear people say, "*Terraform is cloud-agnostic*," it makes my eye twitch. To me, cloud-agnostic would be one configuration file that works in all the major clouds with little to no changes. That isn't true in TF. Using TF for different environments, if that is one of your goals, will require you to get good at working with different HCL resources and their nested key-value configurations.

But even with the need to create environmentally-specific HCL, the fact that the logical process for using TF never changes has a lot of advantages. That, when combined with TF's capabilities in the Google Cloud space, means that TF is a far superior option compared to Google Cloud's Deployment Manager. This is likely why, starting in 2022, Google now recommends TF as the first choice when selecting an IaC tool for use with Google Cloud. Using it will reduce toil, improve consistency, and allow you to build security and infrastructure best practices right into your resource hierarchy. So, let's get TF working for us.

Terraform – the least you need to know

TF is a command-line utility written in Go that can be downloaded from the HashiCorp website. Since it's commonly used with Google Cloud, you can also find it pre-installed in Cloud Shell. If you'd like to install it yourself on a laptop or VM, please go to `https://learn.hashicorp.com/tutorials/terraform/install-cli`.

To use TF, you must run through five logical steps:

- **Scope**: Identify and plan your Google Cloud infrastructure.
- **Write**: Create the HCL files needed to realize the plan.
- **Init**: Download the requisite plugins and prepare the current folder for executing any TF scripts in it.
- **Plan**: View and approve the changes TF will make to your current infrastructure.
- **Apply/Destroy**: Apply (create/update) or destroy the infrastructure that's specified in the current folder of `.tf` configuration files.

Since these steps are used to create, modify, and destroy infrastructure, you tend to iterate through them multiple times. Putting them into a simplified graphic, they would look like this:

Using Terraform

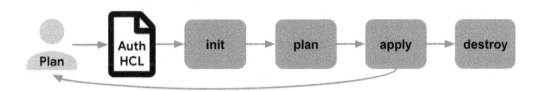

Figure 4.2 – Terraform steps

To help learn how TF works, let's look at an example. We'll start by setting up the recommended TF folder and file structure, with a set of scripts we'll use to create a virtual machine and a Cloud Storage bucket. Then, we'll learn how to use TF to apply the scripts and build the infrastructure with the Terraform CLI. Next, we'll examine the results in the cloud and the state files that TF creates and uses to speed up future infrastructure changes. With some general TF knowledge under our belt, we'll turn back to our foundation laying and create a foundation for future TF usage in GCP, all while following the recommendations laid out in Google's security foundations guide.

Let's get to work.

Setting up an example Terraform module

Imagine that you have a basic **Linux, Apache, MySQL, and PHP (LAMP)** stack application you are planning to build on a VM, and you know it will need to be able to store files in a Cloud Storage bucket. You would like to use TF to create the VM and Cloud Storage bucket. In reality, a lot of ancillary components would need to be added into the mix to make this work at the production level – from a service account to network configurations to firewall rules – but we'll leave them out for now to simplify this example.

> **Note – The Files for This Example Can Be Found on GitHub**
>
> If you check out the GitHub repository at `https://github.com/`
> `PacktPublishing/The-Ultimate-Guide-to-Building-a-`
> `Google-Cloud-Foundation`, you will see the complete and working
> `tf_lamp_fun` subfolder.

To get started, you need to set up the TF folder structure. Though it's possible to create a single folder and do everything within that, even using a single .tf script file, as your infrastructure becomes more and more complex, it's very helpful to divide and conquer with TF modules. A **Terraform module** is a subfolder of TF files (literally, files with a .tf extension) that all work to handle some part of your infrastructure. The root module is either the top-level folder itself or some folder just off the top level (for example, if you wanted to have a development and separate production root module, which we will see later in this chapter). The root module controls the overall process, mostly by putting all the modules into play. Folders that are typically in a nested /modules folder are submodules of that overall process. By default, you should create README.md, main. tf, variables.tf, and outputs.tf text files in each module. *README* is where you document what that module/submodule is doing. *Main* is where you enter the configurations to do the work. *Variables* allows you to define and describe any dynamic variable the script expects, as well as provide default values where appropriate. Finally, *outputs* defines and describes any output values from that particular script.

> **Note – Terraform Doesn't Care about Filenames, It's about the Extension**
>
> While the filenames and purposes I have used in this example are typical, and even a recommended best practice, they are in no way required. TF will evaluate any .tf files it finds in the current folder and process them in the order that TF feels is best. The folder structure and naming conventions are helpful only so far in that they remind you what's defined where. If you feel that it is helpful to subdivide a module by splitting the main.tf file into multiple, smaller .tf files, then that is perfectly acceptable.

So, before I do anything else, let me go ahead and get that initial structure set up. For now, all the files are empty text files. For this example, the folder structure may look as follows. main.tf 1 would be the root script, main.tf 3 would be responsible for creating computing resources (in this example, a VM), and main.tf 2 would handle storage – that is, the Cloud Storage bucket:

Figure 4.3 – Example TF folder and file structure

Let's start with the compute module. Defining resources in Terraform HCL can be accomplished by creating **resource blocks** in your TF files with the following syntax:

```
resource "some_resource_type" "name_for_this_resource" {
    key    = "value"
    key2   = "value"
}
```

Here, some_resource_type is a provider-specific name for what resource is being created, name_for_this_resource will be a reference name created by you and used inside the TF script files for reference, and the key-value pairs are configurations that the resource type expects.

So, if you're going to use TF to create a VM in Google Cloud, you need to check out the documentation for the TF Google Cloud provider for Compute Engine. A TF **provider** is like a driver of sorts, and it provides TF with a set of resources it knows how to implement. We'll see in a bit how the provider gets loaded into the root module, but for now, let's focus on the Google Cloud provider we'll be using to create a VM.

The documentation for the Google Cloud provider can be found here: `https://registry.terraform.io/providers/hashicorp/google/latest/docs`. On the *Google Cloud Platform Provider* home page, down the left-hand side, you will see links to various resource types that it knows how to manage. Expanding **Compute Engine | google_compute_instance** will lead you to the documentation for creating VMs in Google Cloud. There are a lot of options you could specify for a new VM, but many of them have default values. I don't need anything fancy, so I'm going to create my VM like this:

```
resource "google_compute_instance" "app_vm" {
  name                     = "demo-vm"
  machine_type             = var.machine_type
  zone                     = var.zone
  allow_stopping_for_update = true
  boot_disk {
    initialize_params {
      image = "ubuntu-2110-impish-v20220204" //ver 21.10
    }
  }

  network_interface {
    network = "default"
    access_config { //This adds an ephemeral public IP
    }
  }
}
```

This script creates a GCP VM. How can I tell? Because the resource type is set to `google_compute_instance`. The name and boot image values for the VM have also been explicitly declared. The machine type of the VM (chipset, CPU count, and amount of RAM) and the Google Cloud zone where the VM will be created were pulled from variables. The referenced variables (`var.xxx`) should be declared in the `variables.tf` file and their values can be provided in several different ways, as we'll cover soon:

```
variable "zone" {
  description = "Zone for compute resources."
  type        = string
}
```

```
variable "machine_type" {
  description = "GCE machine type."
  type        = string
  default     = "n1-standard-4"
}
```

Anyone using this module can look at my `variables.tf` and learn something about the variables, which ones have default values, and which ones must be assigned a value.

If you'd like to pass anything back from a module to the caller – either something needed as input to some other module (such as a machine IP address) or something you'd simply like to be printed back to the user when they run the script – you can add that to the `outputs.tf` file:

```
output "vm_public_ip" {
//Note: The following line of code should all be on a single
line.
    value = google_compute_instance.app_vm.network_
interface.0.access_config.0.nat_ip
    description = "The public IP for the new VM"
}
```

Once the submodule has been fully defined, it can be called from your root module. Before doing that, though, the top of your root module (`main.tf` 1 in Figure 4.3) should start by loading the provider you are using for this series of scripts. For us, that's Google Cloud's provider, which is documented here: `https://registry.terraform.io/providers/hashicorp/google/latest/docs`. Our root script will start something like this:

```
provider "google" {
  project = var.project_id
}
```

After loading the provider, a typical root module will call all the child modules, pass them any variable values they require, and let them do the majority – if not all – of the work. So, the rest of my `main.tf` file might look like this:

```
module "compute" {
  source = "./modules/compute"
  zone   = var.zone
  depends_on = [
```

```
        module.storage
    ]
}


module "storage" {
    source     = "./modules/storage"
    region     = var.region
    project_id = var.project_id
}
```

Notice that for compute, I'm specifying the VM's zone using another variable, but I left out the machine type because I was happy with its default. I also wanted to make sure the GCS bucket was defined before the VM, so I added depends_on to the VM, which tells TF to make sure the storage module has been fully created before processing compute. TF can typically determine a good order of creation based on physical dependencies between resources, but if there is a dependency that TF can't glean from the script, then this is a way to force ordering.

Since I passed in values to the modules that, once again, came from variables, I must declare or re-declare those variables in the variables.tf file for the root module, like so:

```
variable "project_id" {
    description = "Project owning defined resources"
    type        = string
}
variable "zone" {
    description = "The default GCP zone"
    type        = string
}
variable "region" {
    description = "The default GCP region"
    type        = string
}
```

Let's get back to the variables. There are several ways TF variable values can be provided when executing your scripts, including the following:

- Via command-line prompts when the script is applied with terraform apply (the default). Upon running the script, TF sees a variable missing a value, and asks you for it then and there.

- By storing the values in environmental variables on the machine where you are executing the script using `TF_VAR_your_variable_name`, like so:

  ```
  export TF_VAR_zone=us-central1-a
  ```

- By passing the values in with the `apply` command, like so:

  ```
  terraform apply -var="zone=us-central1-a"
  ```

- By using a `.tfvars` file

I don't want to have to specify the variables manually, so I'm going to create a `.tfvars` file:

```
project_id = "patrick-haggerty"
region="us-central1"
zone="us-central1-a"
```

With all the files in place, including the Cloud Storage files I've omitted from this book, let's create our infrastructure.

Using Terraform

Even though your TF module structure and files are in place, there are several key TF CLI commands you will need to continue. These commands will all be executed from within the root module folder:

- `terraform init`: Download any provider files needed to implement the current TF modules.

- `terraform validate`: Check the TF configurations for validity.

- `terraform plan`: Generate a human-readable plan of what the application of the current configurations would do to the infrastructure.

- `terraform apply`: Implement or update the infrastructure defined in the TF scripts.

- `terraform destroy`: Remove any infrastructure created by the current scripts.

To implement my example, I'm running my scripts out of Cloud Shell, which comes with the TF CLI pre-installed. To use the CLI, I change to the folder containing my root module. Here, that would be the `tf_lamp_fun` folder. First, I must run `init` to download all the requisite provider files, and then run `validate` to sanity check the various TF files throughout the module structure.

When you think everything is ready to roll, you can use `terraform plan`. A hierarchical view of exactly what the scripts, when applied, will do will be returned. It will include several common symbols, including the following:

- `+`: Create
- `-`: Destroy
- `-/+`: Replace by destroying and re-creating
- `~`: Update in place

This is your chance to examine what TF is about to do. You'll notice that the resources display the values for all the properties – the ones you explicitly assigned and the ones with defaults. If the plan appears to be doing what you like, then `terraform apply` it.

A note on Terraform state

If you create and run a series of TF scripts from the command line, you'll notice that several files are created:

- `terraform.tfstate`: A JSON file of what was implemented by TF in your environment the last time the modules were applied with `terraform apply`.
- `terraform.tfstate.backup`: A backup copy, one generation back, of what was applied by TF.
- `.terraform.lock.hcl`: Version information for modules and providers.

By default, each time `terraform plan/apply` is executed, a refresh will be completed of the state file from the live infrastructure. This ensures that no changes have been made behind TF's back. As a best practice, all infrastructure changes should be made through TF, making this an unneeded and potentially time-consuming process. Adding `-refresh=false` will disable this functionality.

Two additional TF CLI commands related to state are as follows:

- `terraform output`: Shows any output values from the last `terraform apply` command.
- `terraform show`: Shows the contents of the last state snapshot in full.

Remember that TF isn't just used to create and destroy infrastructure – it's also used to update it. After some testing, I've decided that the default n1-standard-4 machine type for my VM is too big, wasting both resources and money. I'd like to change it down to n1-standard-1. The compute module already has the variable set up for machine_type, but the root module doesn't. One option would be to update the variables.tf and terraform.tfvars files, just like we did with our other variables, but I'd like to show a slightly different approach this time. I'm going to go to the top of my root main.tf file and, just below provider, I'm going to add some locals. locals are used by Terraform to create and define variables directly in a specific .tf file:

```
locals{
    cpu_count = "4"
    chip_type = "standard"
    chip_family = "n1"
    machine_type = format("%s-%s-%s", local.chip_family,
  local.chip_type, local.cpu_count)
}
```

If you're thinking, "*Hey, we could use locals to define all the variables,*" you could, but I wouldn't. This approach tends to work best when you use the .tfvars file for key naming elements, such as base names, and then use tricks such as the format() function, to combine the base elements into variable values. Say you're creating a bucket, and your buckets all use the bkt-business_name-app_name-#### format, where the business name and app name could be defined in .tfvars and then used as name parts for multiple resources.

To use the local variables I created here, I could update the compute module's details, like so:

```
module "compute" {
  source         = "./modules/compute"
  zone           = var.zone
  machine_type = local.machine_type //notice the local.
  depends_on = [
    module.storage
  ]
}
```

When I run `terraform plan`, I will see the following:

```
~ resource "google_compute_instance" "app_vm" {
    ...
        ~ machine_type = "n1-standard-4" -> "n1-standard-1"
```

Note ~, indicating that the resource will be updated in place. So, TF is going to power the machine down, edit its machine type specification, and then restart it. Nice.

> **Note – Terraform Has a Lot of Useful Functions**
>
> You can find them documented at `https://www.terraform.io/ language/functions`.

Laying a foundation for Terraform use in Google Cloud

Running TF scripts out of Cloud Shell, as I did for my demo, is nice and easy, but it will quickly become problematic if you try to use that same approach to manage an organization full of infrastructure. Some such problems are as follows:

- Terraforming infrastructure, when done right, starts with you creating the GCP resource folders and projects themselves (which we will be doing soon). This means there needs to be a central location (project) from where all TF scripts are run.

- The TF scripts themselves should be stored and versioned, and infrastructure changes should be applied automatically. This implies the usage of an automated CI/CD pipeline, starting with a Git repository containing the scripts, and ending with the implemented changes.

- The power to create and manage all infrastructure across the entire organization would require a very high level of access for the person running the script. A better way to do this would be to run TF under some high-powered service account and then control access at the entrance of the CI/CD pipeline.

- The need to manage infrastructure will be bigger than the need to manage an individual, so besides a central Git repository driving the automated Terraform CI/CD pipeline, a secure storage location for the state files, which frequently contain sensitive information, will also be required.

Since Google has switched its recommended IaC automation tool from Deployment Manager to TF, they have also released some foundational best practices in a PDF with corresponding TF scripts. Many of the best practices have been built into this book, but you can check out Google's latest security foundations guide at `https://services.google.com/fh/files/misc/google-cloud-security-foundations-guide.pdf`. The corresponding TF scripts can be found at `https://github.com/terraform-google-modules/terraform-example-foundation`. They are, in turn, based on Google's more general examples of Terraform scripts, which can be found in their Cloud Foundation Toolkit at `https://cloud.google.com/docs/terraform/blueprints/terraform-blueprints`.

You don't have to use all of Google's TF scripts as-is, but they can make a good foundation for scripts you create on your own.

To get a solid foundation in place where you can use TF to manage the infrastructure across your entire Google Cloud footprint, you should start by creating two projects and placing them in a GCP folder toward the very top of your organizational structure: `seed` and `cicd`. Why two? Mostly because it keeps things very clean from a **separation of concerns (SoC)** point of view. CI/CD has its own set of requirements, so let's package them all into one project, while everything TF lives in the other:

- `cicd`: This hosts the CI/CD pipeline itself, along with the Git repositories containing the TF scripts. If you are using Jenkins as a CI/CD tool, its build servers will live here. CI/CD will also host a pipeline service account, which will impersonate a high-level `seed` project service account. This, in turn, will contain all the requisite access to make infrastructure changes. Access to this project, as well as the various Git repositories in it, will be granted to personnel that need to manage infrastructure.

- `seed`: This is where the shared TF state file will be encrypted and stored, and where the highly privileged service account that will be used to make infrastructure changes will live. Access to this project will be highly privileged.

The easiest way to implement these two projects is to use the first part (0-bootstrap) of Google's Example Foundation at `https://github.com/terraform-google-modules/terraform-example-foundation/`. This repository was built to implement an entire foundation layer that follows the Google Security Best Practice Guide, which can be found at `https://cloud.google.com/docs/terraform/blueprints/terraform-blueprints`. It's constructed from other Google TF blueprints, including `bootstrap`, `folders`, and `project-factory`. If the full sample org scripts don't work for you, see if you can mix and match some of the others to get the work done the way you need.

The Example Foundation scripts can use Google's Cloud Build or OSS Jenkins to do the CI/CD build work. To keep things simple, I'm going to stick with the default: Cloud Build.

To start, you will need an existing project in your organization that you can use as a launching pad. Alternatively, you could perform these steps using the Google Cloud SDK on your laptop. You will also need to log in to Google using your organizational administrator account. If you are using my naming scheme, then this will be in `gcp-orgadmin-<first>.<last>@<domain>` format.

Next, you must fork Google's sample foundation repository of scripts so that you can edit them as you need. I've done exactly that, and I've put my copy in with the rest of the code from this book, so you can find it as a `terraform-example-foundation` sub-folder in my repository at `https://github.com/PacktPublishing/The-Ultimate-Guide-to-Building-a-Google-Cloud-Foundation`.

> **Note – You Should Use the Latest Version of the Example Foundation**
>
> I'm currently working with the 2021-12 version of Google's blueprint guide and the 2022-06 latest version of the TF scripts. I don't doubt that Google will be updating these files and that as they do, the steps I'm putting in this book – the steps that work with my fork of Google's scripts – may start to drift from what you are seeing with the latest version. Please check the main repository and its documentation for updates.

A note on labels. In Google Cloud, a label is an extra identifier that helps you track resources at a finer grain. I'll talk more about labels later in this chapter, so for now, when I have you editing labels, think of using them to identify data that's been placed on sticky notes and stuck to the side of various resources. Both the label key and its value must contain UTF-8 lowercase letters, numeric characters, underscores, and dashes.

Either on your laptop or in your launching pad project's Cloud Shell window, clone down your copy of Google's Cloud Foundation Toolkit scripts. Here's an example where I'm using my copy. Again, this should be updated to reflect your fork of Google's scripts:

```
git clone \
https://github.com/PacktPublishing/The-Ultimate-Guide-to-
Building-a-Google-Cloud-Foundation
cd The-Ultimate-Guide-to-Building-a-Google-Cloud-Foundation/
chapter04/terraform-example-foundation
```

Now that you have downloaded the scripts, let's make some variable and value edits:

1. Change to the 0-bootstrap subfolder.

2. Take a minute to examine the documentation for the possible variables that the bootstrap TF scripts will accept by examining the variables.tf file.

3. Copy the terraform.example.tfvars file and change its name to terraform.tfvars. As a side note, it's not uncommon to use the provided .example file as a starting point for new .tfvars files.

4. Edit your terraform.tfvars file. Insert values for your org_id and billing_account ID. These can be retrieved using the console, or by running the following commands:

    ```
    gcloud organizations list
    gcloud beta billing accounts list
    ```

5. Next, edit the group_org_admins and group_billing_admins security group names. If you have been following along and using mine and Google's recommended names, then all you should need to do is change the domain names.

6. Lastly, edit default_region and set it to whichever region would work best for you. By default, any region-specific resources that are created by the sample TF scripts will be physically put here. It should be a region with BigQuery support.

7. Check the documentation for the 0-bootstrap TF module and make sure you don't need to make any other variable changes. Save and close the terraform.tfvars file.

8. Open the root module's main.tf file from the root of the 0-bootstrap folder. Find the two project_labels blocks. These two blocks are going to assign some identifying labels (sticky notes) to the seed and cicd projects. Feel free to use whatever values make sense to you. If you use internal billing codes to track spending (remember that this will be a shared-use project), change the two billing_code values as appropriate.

9. Next, add a primary and, optionally, secondary contact. If there are questions about these projects, it's these people who should be contacted. Remember the labeling naming requirements. I typically use something in the form of first-last (patrick-haggerty), though a team name may work even better than an individual here (gcp-coe-team).

10. The business code should uniquely identify the business unit or group within your organization that's responsible for these projects. Since these projects don't belong to a business unit, as they will be commonly used, I'm going to use `zzzz` to denote that.

11. Make sure both of your `project_labels` sections have been updated, then save and close the file.

Excellent! Now that changes have been made to the variables, let's use the scripts. Again, if you are using a newer version of Google's example scripts, double-check their latest documentation and make sure these steps haven't changed:

1. Open a command-line terminal and change to the `0-bootstrap` subfolder.

2. Use `terraform init` to load all the required TF provider and remote module files.

3. Use `terraform plan` and review the output, which will be quite extensive.

4. `terraform apply` the scripts. It will take several minutes for them to run.

5. Use `terraform output terraform_service_account` to retrieve the service account that TF will run under. Put this value in a note file as you will need it in a later step.

6. Use `terraform output gcs_bucket_tfstate` to grab the GCS bucket where the state file will be stored. Add this to your notes from the preceding step.

7. Use `terraform output cloudbuild_project_id` to get the project ID for the `cicd` project. Add this to your notes.

8. Make a copy of the `backend.tf.example` file and name it `backend.tf`.

9. Open the `backend.tf` file and replace `UPDATE_ME` with the name of the TF state bucket you copied a couple of steps back.

10. Rerun `terraform init`. The TF backend resource type can be used to define a remote storage location for the TF state file. Re-running `init` will move the state file from the directory where you are applying the commands to a newly created Cloud Storage bucket in the `seed` project.

That should be it. If you look at the resource hierarchy for your organization in the GCP Console, by going to **Navigation menu | IAM & Admin | Manage Resources**, you should see a new `fldr-bootstrap` folder with the two new projects under it – that is, `proj-b-cicd` and `proj-b-seed`. You will also see the project labels we applied. This will look similar to the following:

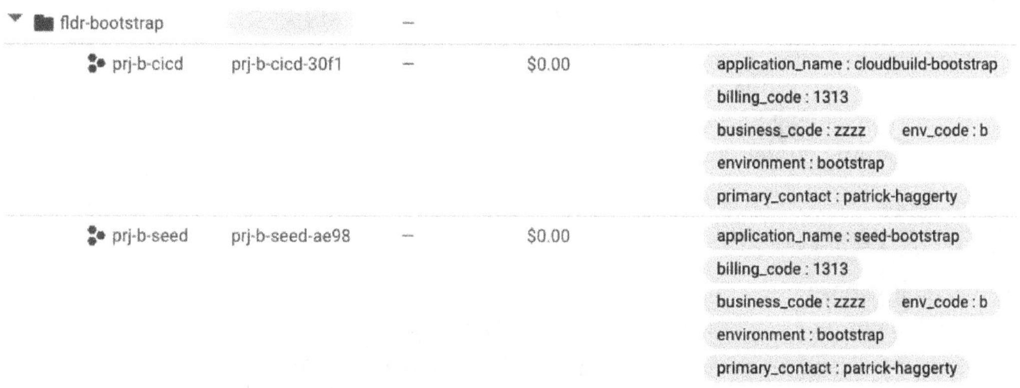

▼ ▪ fldr-bootstrap		–		
▪ prj-b-cicd	prj-b-cicd-30f1	–	$0.00	application_name : cloudbuild-bootstrap
				billing_code : 1313
				business_code : zzzz env_code : b
				environment : bootstrap
				primary_contact : patrick-haggerty
▪ prj-b-seed	prj-b-seed-ae98	–	$0.00	application_name : seed-bootstrap
				billing_code : 1313
				business_code : zzzz env_code : b
				environment : bootstrap
				primary_contact : patrick-haggerty

Figure 4.4 – Resource Manager view of new projects

Remember, the `seed` project will be doing the work, but next to no one should have direct access to it. The people administering your organizational structure will be doing that from the `cicd` project.

If you use the project selector (remember, that's a little to the right of the Navigation menu's hamburger button), you can navigate to your `prj-b-cicd` project. Once there, look at the project's source repositories. You will see a series of Git repositories that have been created by the bootstrap process. Once you finish this book and have a better grasp of how this structure works, you will need to consider if this exact set of repositories makes sense in your organization, or if you will need something different. The list of repositories that have been created by the bootstrap TF scripts is one of its variables, so adding, removing, or renaming repositories will be easy to do, thanks to TF.

We will work with these various repositories and see how the CI/CD pipeline works shortly. First, though, it's time to get to the fifth major step in Google's 10 steps of foundation laying and start building our resource hierarchy.

Step 5 – creating a resource hierarchy to control logical organization

Do you remember how we started this chapter? With figuring out where everyone got to sit in the new office building? It's time to do exactly that, only for Google Cloud.

A well-planned Google Cloud resource hierarchy can help the entire GCP footprint by implementing key configuration and security principles in an automated way. If you've set up your bootstrap infrastructure, then you will have the CI/CD-driven TF projects in place, so let's start laying in other key folders and projects. To do that, we'll start by laying down some naming guidelines.

Naming resources

It is doable, if you want to willy-nilly name your GCP folders and projects, but when you're implementing a level of automation, consistency in naming makes sense. Do you remember the document we created back in *Chapter 2, IAM, Users, Groups, and Admin Access*, where we put the names for things such as our organizational administrator email address format? It's time to break that out again, or time to create a new infrastructure naming document.

One of the standard naming features you may or may not decide to implement is Hungarian Notation. Throughout grade and high school, I attended an all-boys Catholic school in Irving, Texas named Cistercian. Most of the Cistercian monks at that time had escaped Hungry in or around the 1956 Hungarian Revolution, when Hungry temporarily threw out the Stalinist USSR government they'd been under since the end of WWII. Hungarians traditionally reverse their name, placing their family name before their given name. So, Hungarian-style naming would change my name from Patrick Haggerty to Haggerty Patrick.

In the 1970s, a Hungarian named Charles Simonyi, a programmer who started at Xerox PARC and later became the Chief Architect at Microsoft, came up with what we now call Hungarian Notation. Hungarian Notation advocates starting names with a prefix that indicates what exactly is being named. For this book, I'm going to use Google's naming recommendations, which will leverage a type of Hungarian Notation. You can decide if Google's naming style works for you, or if you'd like to implement naming some other way.

To see the naming guidelines from Google, check out their Security Foundations guide at `https://services.google.com/fh/files/misc/google-cloud-security-foundations-guide.pdf`. Let's start with the names we are going to need to create an initial resource hierarchy. Note that {`something`} implies `something` is optional:

Resource	Naming Convention
Folder	`fldr-environment` Example: `fldr-prod`
Project	`prj-business-code-environment-code{-label}-unique-number` Example: `prj-dep4-p-converter-app-12345`

Following Google's recommendation, notice that I'm starting my names with a Hungarian Notation that identifies what is being named, such as `fldr` for `folder`. Then, I'm using a mix of standard and user-selected naming components. The following is for reference purposes:

- `environment`: A folder-level resource within the organization. Values include `bootstrap`, `common`, `prod`, `non-prod`, and `dev`.

- `environment-code`: A short form of the environment name. Values include b, c, p, n, and d.

- `label`: Descriptive element to enhance the name.

Now that we have an idea of how we are going to name things, let's discuss how we will lay out our resource hierarchy.

Designing the resource hierarchy

There are lots of ways your resource hierarchy can be laid out, and many would be perfectly acceptable. Instead of looking at lots of examples, let's look at two good and one not-so-good designs. First, let me show you a very common design that is, unfortunately, what you shouldn't do:

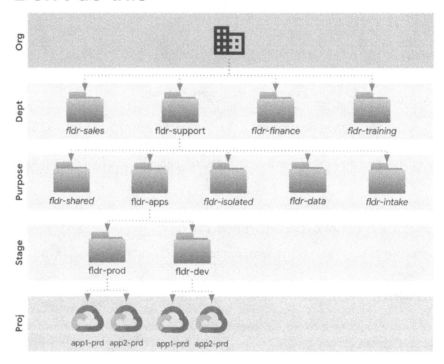

Figure 4.5 – Bad resource hierarchy example

I can hear some of you now. "*What? This is the bad example? But this is exactly what I was planning on doing!*"

To get this design, I imagined looking at a company org chart, and I created a resource hierarchy that matched it. On the surface, this seems reasonable (which is why I see it all the time, unfortunately). So, what's wrong with it? Let's take a look:

- Org charts change, so how would you implement those changes? And if you ignore them, how soon before the logical structure has no relationship with the org chart?

- There's no place for shared resources. So, where would I put the `seed` and `cicd` projects? How about a shared VPC or the VPN connection to our on-premises data center?

- Security might be tough to implement. Does each of the departments have internal developers? If so, there's no easy way to control their access, other than doing it project by project.

- How could you easily apply an organizational policy to all the development environments while forbidding them from having external IPs? Here, you'd have to do it department by department or something.

> **Note – Folders Have Limits**
>
> Whichever design you decide on, keep in mind that there are some limits with folders. Currently, Google limits folder depth to no more than 10 levels, and any given folder can have up to 300 direct subfolders.

OK; if the org chart approach is wrong, what might be better?

Well, let's start with a few Google Cloud principles and behaviors:

- Folders logically divide projects, which, in turn, wrap resources and attach billing.

- **Organizational policies** allow organization policy administrators to define guardrails that limit particular types of resource configurations. For example, an org policy could be created to forbid Compute Engine VMs from having public IP addresses or to ensure that data is only stored in Cloud Storage buckets in a particular part of the world. Exceptions to policies may be made for particular subfolders or projects, but the ability to override policies requires a high level of access.

- IAM policy bindings allow permissions to be assigned to users or groups at any point, from top to bottom, from organization to resource. However, permissions assigned at a higher level, such as a folder, may not be revoked at a lower level, such as a project. We will discuss this more in *Chapter 5, Controlling Access with IAM Roles*.

Google is a big fan of a basic folder structure, just enough to simplify policy application, and filled with lots of purpose-built projects that follow a consistent naming convention. I think the recommendation makes sense. Have you ever heard of the **Keep It Simple Stupid (KISS)** principle? You should apply it to your resource hierarchy design. Imagine your organization and what it does, then split that purpose along the most major lines, not department lines, but more *"what can Google do for you"* lines.

It might be because of my development background and my introduction to Google Cloud from that perspective, but I like the basic approach Google takes to its security blueprint guide. This is a design focused on the belief that you're using Google primarily as a place to build and run applications. It's simple but easily extensible. It may look like this:

By environment design

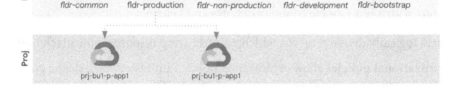

Figure 4.6 – By environment design

See? Simple, right? All the shared resources sit in a common folder and the application-related projects all sit in folders related to their stage of development, ending with the bootstrap folder that hosts the `cicd` and `seed` projects. The layout allows us to assign base permissions and organizational policies at a very high, centralized folder level, and we can add any projects we need while following a pre-agreed-on naming convention.

While the design is nice from an application development and deployment view of the world, it doesn't address needs around centralized data storage, regulatory isolation requirements, or multiple departments doing development. The fix? Add a few more folders.

If your operating units have a lot of autonomy, then you may wish to consider adding a layer of business unit folders between the org and env folders. This would allow the EMEA business unit to have their copy of the environments, and then the AMER business unit could do the same. If you have multiple development teams, then you could add team folders just inside each environment folder. If you have centralized data and data warehousing needs, then a folder for that can either be in the common folder, or you could add a new top-level folder and put those things there. In the end, including our folder naming convention, the design might be closer to this:

By business unit, environment, and team

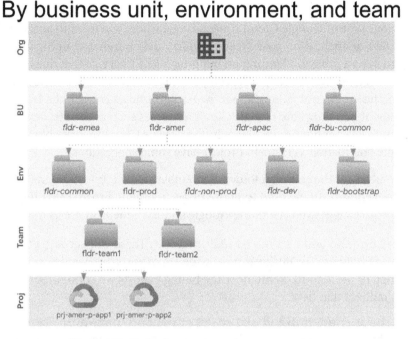

Figure 4.7 – By business unit, environment, and team

To fully flush out your design, but you should also consider things such as the following:

- Do you want to leverage the folder approach for separating by business unit, or do you want the business units to share the same folders and then use project naming to identify which project belongs to which business unit?

- In Google Cloud, an organization can have sub-organizations that are completely isolated, with different users, projects, and policies – everything (`https://cloud.google.com/resource-manager/docs/managing-multiple-orgs`). Does your business need this extreme level of isolation? This change would eliminate my business unit folders and replace them with multiple organizations.

- Do you need to comply with any regulatory requirements? If so, how can projects and folders help? (`https://cloud.google.com/security/compliance`)

- Do the bootstrap projects belong at the highest level of the organizational structure, inside the business unit, or both?

But your planning won't stop with folders – you will also likely need some common-use projects to go along with them.

Planning common projects

The most obvious type of Google Cloud project is one where you need to accomplish some specific task or application goal. Need to build a data warehouse in Big Query? You're going to need a project. Planning on building a REST service that does currency conversion? You'll need a dev/non-prod/prod project for that too. But what about projects for background tasks? I'm not talking about the background as in behind a UI – I'm talking more about handling core Google Cloud resources such as secrets, centralized networking, and DNS configurations. Luckily, the Google Cloud security blueprint and its related Example Foundation Terraform project have some suggestions.

Let's start with ideas for the common folder. Remember that `fldr-common` (don't forget our naming convention) will be used to house projects serving broad, cross-department needs. Some project suggestions from the Google security blueprint are as follows:

- `prj-c-logging` and `prj-c-billing-logs`: These will act as centralized locations to hold org (or business unit) wide log exports, with auditing in the first and billing in the second. Remember the billing exports we configured? We would need to redirect that here.

- `prj-c-base-net-hub` and `prj-c-restricted-net-hub`: These will hold the hubs, if you're using a hub and spoke network design. This will allow environment-specific shared VPCs to route cross-environment traffic and/or traffic that needs to rout out through Interconnect.

- `prj-c-dns-hub`: This is a central project where all the org/business-wide DNS configurations will reside.

- `prj-c-secrets`: This is for org-wide secrets and encryption keys.

- `prj-c-interconnect`: This holds the GCP end of your Interconnect (fiber) connection back to your on-premises data center.

- `prj-c-scc`: This is the central location for managing Security Command Center alerting.

Once you have your common folders planned out, think about projects that you may need in every environment. If you are using the dev/non-prod/prod design, then you may want to pre-construct projects that can centralize resources at the environment level. Some examples are as follows:

- `prj-d/n/p-monitoring`: This hosts cross-project metric scopes, for any monitoring that needs to happen across the entire environment.

- `prj-d/n/p-secrets`: This contains environment-wide secrets and keys.

- `prj-d/n/p-shared-base` and `prj-d/n/p-shared-restricted`: These are for environment-specific shared VPC networks. These may be linked to hubs in the common folder, depending on your design.

While these are two nice starter lists, they are in no way meant to be exhaustive or mandatory. You may not be using Interconnect, so you wouldn't need that project. You also may not want a way to cross-share VPC traffic, in which case the hubs would be unnecessary. And there's no telling what other centralized or environmental services you may need that are specific to your organization, so make sure that you tweak the design to something that makes sense to your organization.

Before we move on and apply our infrastructure, let's talk a little about resource labeling.

Labeling resources in Google Cloud

To help you manage organizations as they grow, you should seriously consider coming up with and using a core set of labels. As I mentioned earlier, in Google Cloud, a label is an arbitrary *key-value* piece of metadata you can stick to the side of most GCP resources. They can be used for tracking spend at a more granular level, and to help better filter GCP logs, among other things.

Common labels include things such as team, owner, region, business unit, department, data classification, cost center, component, environment, and state. We are going to apply a nice standard set of labels to our projects, starting with the following:

- `environment`: dev/nonprod/prod/common/bootstrap.
- `application_name`: Exactly what it says.
- `billing_code`: An internal billing code. It's up to you how you implement it.
- `primary` and `secondary_contact`: A contact person or preferably group.
- `business_code`: A four-character code that describes the business unit. Again, make of it what you will.
- `env_code`: d/n/p/c/b: Our aforementioned environments.

For more information about labels, go to `https://cloud.google.com/resource-manager/docs/creating-managing-labels`. As a best practice, using a labeling tool or applying labels via automation such as TF, is advisable.

Implementing a resource hierarchy

To implement my resource hierarchy, I'm going to continue using Google's Example Foundation, which is, in turn, built on several of their other blueprints. All the blueprints can be found at `https://cloud.google.com/docs/terraform/blueprints/terraform-blueprints`. The ones you should pay most attention to at this point are:

- The *Example Foundation*, which, as I've said, is built by combining and modifying several of the other blueprints.

- The `bootstrap` module, which we've already used as part of the Example Foundation to create our CI/CD and `seed` projects.

- The `folders` module, which can create multiple folders under the same parent, assigning per folder and all folder admins.

- The `project-factory`, which we will use later to build projects with a default set of network configurations, IAM policies, service accounts, and enabled APIs.

The resource hierarchy I'm going to implement at this point is the basic five-folder version I first showed you as a form of good design. It would be trivial to use the `folders` module to add the extra business unit and team folders if I decided to go that route.

One thing to keep in mind is that your folder design isn't set in stone. It is possible to move projects from one folder to another, so it's possible to start with one design and decide to modify it. Having said that, it's a lot easier to begin with a simple design that works and then grow from there, thus minimizing the need to move things around.

> **Note – Moving Projects Will Have IAM Repercussions**
>
> If you move a project from one point in the resource hierarchy to another, any IAM configurations you've directly set inside the project will move with it. However, the inherited permissions will change based on the project's new location in the hierarchy. For more information on moving projects, go to `https://cloud.google.com/resource-manager/docs/moving-projects-folders`.

Before we dive deeper into creating our folder structure, let's talk about how the `cicd` and `seed` projects work.

Using the cicd and seed projects

When we created the `cicd` and `seed` projects, we didn't talk about how to use them. As you may recall, the `seed` project is where the high-powered TF service account lives, and where the TF state information is stored and encrypted. The `cicd` project is where infrastructure managers will work to submit changes to the CI/CD pipeline.

> **Note – You May Not Have the Project Quota You Need**
>
> The following example creates a passel of projects. As a result, Google may throw an error related to your TF service account, essentially saying that the user doesn't have enough of a quota to create and set up billing on so many projects. You should go ahead and ask Google to increase the SA's quota. Go to `https://support.google.com/code/contact/billing_quota_increase` and request a quota increase of 50 for the TF service account you generated while creating your bootstrap project. This should be in your notes.

If you navigate to your `cicd` project (`proj-b-cicd`) and then go to **Navigation menu | Source Repositories**, you will see that you currently have several Git repositories. The `0-bootstap` module created them as part of its setup task, and the actual repository list was controlled by the `cloud_source_repos` TF variable. The current set of repositories includes `gcp-environments`, `gcp-networks`, `gcp-org`, `gcp-policies`, and `gcp-projects`. We will cover the purposes of these different repositories soon.

Next, if you return to the cloud console and navigate to **Navigation Menu | Cloud Build | Triggers**, you will see that there are several pairs of Cloud Build triggers attached to these same repositories. Cloud Build triggers watch for repository changes and respond by following a prescribed set of actions defined in a YAML file.

The CI/CD pipeline is going to integrate with Git via a persistent branch strategy. That is, we are going to create and use Git branches to differentiate stages, such as dev, non-prod, and prod, and split between planning and applying TF scripts.

So, the plan/apply triggers work against the dev, non-prod, and prod environments. If you want to plan a job in the production environment, you would need to do the following:

1. Edit your various TF-related files.
2. Create a `plan` branch in Git, add your changed files to it, and push it to your repository up in GCP. This will trigger a plan build, which you can inspect.

3. Now, you must decide which environment you want to push your changes to. Let's say it's production. Here, you would switch to a `production` branch created from the same `plan` branch you just pushed to your CI/CD project.

4. When you push the `production` branch to `origin`, up to GCP, it will trigger `terraform apply` and run it against your production resources.

Not too shabby. It makes good, logical sense and is easy to use.

To understand this process, let's implement the `1-org` and `2-environments` parts of the Example Foundation.

Creating our organizational base folders

The next step in Google's TF Example Foundation is `1-org`. It sets up several projects in the common folder for things such as org-level secrets, central networking, and org-level logging. It also configures a slew of base security features. Most of this step deals with things we will discuss later in this book (access, security, and logging) since Google's 10-step Cloud Foundation deals with them in a slightly different order.

Though we don't get into the details of exactly what `1-org` does at this point, it will nicely illustrate how our CI/CD pipeline works.

To check out the latest instructions for `1-org`, go to `https://github.com/terraform-google-modules/terraform-example-foundation/tree/master/1-org`.

To start, make sure you are logged into your organizational administrator account. Then, either open Cloud Shell in the same launchpad project you used to deploy the bootstrap project or go back to the terminal window on your laptop if that's where you deployed from. You will need to locate the folder where you cloned down the Example Foundation. Some of the steps we will follow use the new beta TF validator to check some of the things we will do against a set of policy configurations. This is not a required component, but one that the Example Foundation uses, so I'm leaving it in.

Let's get started by loading the Terraform validator policies into the related Git repository:

1. In Cloud Shell or your laptop terminal window, change to the same top-level folder where you cloned `terraform-example-foundation`, but don't change to the Example Foundation folder itself.

2. In the text file containing the notes that you started when creating the bootstrap projects, retrieve the project ID of the CI/CD project. If you didn't grab the name, you can also look at your resource hierarchy by going to **IAM & Admin | Manage Resources**. Locate the CI/CD project and copy the ID from the ID column.

3. Clone the `gcp-policies` repository you created in your CI/CD project as part of building the bootstrap projects. Make sure that you insert the ID of your CI/CD project:

```
gcloud source repos clone gcp-policies \
--project=CICD_PROJECT_ID
```

4. Change to the cloned empty folder:

```
cd gcp-policies
```

5. Copy the contents of the provided example Terraform policy library into the new folder:

```
cp -RT ../terraform-example-foundation/policy-library/ .
```

6. Commit the changes and push your `policies` master branch into your CI/CD project:

```
git add .
git commit -m 'Initial commit, sample policies'
git push --set-upstream origin master
```

With the policies where we need them, let's get a copy of Google's base example organization in place. This will create the `common` folder and the central use projects we mentioned earlier in this chapter. Before we apply it, though, let's get the files ready:

1. Navigate out of your `policies` repository and back into the main working folder by using `cd ...`

2. Clone down your empty `gcp-org` repository. If you get any warnings about the repository being empty, ignore them:

```
gcloud source repos clone gcp-org \
--project=CICD_PROJECT_ID
```

3. Change into the newly created `gcp-org` folder, then create and check out a `plan` branch. Remember that when we push the `plan` branch up to our source repository in our CI/CD project, it does a `terraform plan`:

```
cd gcp-org
git checkout -b plan
```

4. Copy the example `1-org` folder we cloned from Google into your `gcp-org` folder:

    ```
    cp -RT ../terraform-example-foundation/1-org/ .
    ```

5. The `plan` and `apply` triggers that we created when setting up the bootstrap project expect us to provide Cloud Build YAML configuration files. Let's copy in Google's starter YAML files:

    ```
    cp ../terraform-example-foundation/build/cloudbuild-tf-*
    .
    ```

6. Also, copy in the wrapper script and make it executable:

    ```
    cp ../terraform-example-foundation/build/tf-wrapper.sh .
    chmod 755 ./tf-wrapper.sh
    ```

7. One of the things this example script does is set up an Access Context Manager Policy to allow access from particular IP ranges. First, though, you'll need to check if you have an existing ACM policy. You will need your organization ID for this command, which you can find next to the root organizational node by going to **Navigation Menu | IAM & Admin | Manage Resources**. If you receive a message about needing to enable an API, enable it, wait a minute, and try the command again. If no value is returned, then you're fine. If you get a policy name back, record its name in your notes for use later:

    ```
    gcloud access-context-manager policies list \
    --organization YOUR_ORGANIZATION_ID \
    --format="value(name)"
    ```

8. Make a copy of the `example.tfvars` file so that you can add your configurations:

    ```
    cp ./envs/shared/terraform.example.tfvars \ ./envs/
    shared/terraform.tfvars
    ```

    Excellent! At this point, we have everything prepped. Let's examine what we have and make a few configuration changes before we apply them.

9. Open an editor of your choice to explore and, where needed, modify the files in the `gcp-org` folder. If you are in Cloud Shell, Cloud Shell Editor will work well. If not, anything from VI to Visual Studio Code should work.

10. In the editor, navigate to the `gcp-org` folder and open the `plan` and `apply` YAML files.

Here, you will see a pair of basic Cloud Build configuration files. Cloud Build, as I mentioned earlier, is a serverless Google Cloud CI/CD tool. To tell it what you want it to do, you must create a configuration file, similar to the two we are examining. These scripts will decide whether to use your account or to use a central service account (we will be using the SA), drop the name of our central TF state bucket in where needed, and then call a bash script file to do the TF work. Here, you can see how the bash scripts (`entrypoint: /bin/bash`) are called and the arguments that are passed to each.

11. In your editor, open the `tf-wrapper.sh` file.

The script contains wrappers for all the TF steps, from `init` through `apply`, passing in appropriate arguments where needed.

Next, if you look in the base of your `gcp-org` folder, you'll see a subfolder named `shared`. Expanding it, you will see all the `.tf` files Google has created to do the work. Remember how TF works – any file ending in `.tf` will be applied. In this case, instead of a `main.tf` file that calls a bunch of modules, the designers simply broke the process down into a single folder containing lots of `.tf` files. We've seen some of these files previously (`*.tfvars`, `backend.tf`, and so on). Some of the others you should note are as follows:

- `folders.tf`: This uses the `folders` blueprint to create a `fldr-common` folder under the specified parent, which is pulled from a TF variable. In my example, this will be directly under my organizational root.

- `iam.tf`: This makes a slew of IAM configuration changes, including setting up permissions for various audit and billing log exports on those projects, setting permissions for billing log exports, configuring billing viewers, security reviewers, audit log viewers, and more. Essentially, if any of our common projects need particular access to be configured, it's done here.

- `log_sinks.tf`: This sends a copy of all the logs for all the projects to a storage bucket. In addition, it configures sinks for all Google Cloud audit logs (which we will discuss later in this book) to BigQuery and Pub/Sub. If you'd like to disable any of these sinks, simply comment out the related section.

- `org_policy.tf`: This sets a host of organizational policies, which we will discuss more later in this book.

- `projects.tf`: This uses the Google TF project factory project to create the projects in the `common` folder.

- `scc_notification.tf`: This configures a **Security Command Center** (**SCC**) alert to a pub/sub topic, along with the related subscription, for all SCC notifications.

Now that we've taken our tour, let's tweak a few things:

1. In your editor, open the `projects.tf` file.

2. Search for labels and update the GCP labels that are being applied to each of the common folder projects. The `environment`, `application_names`, and `env_code` labels are probably fine, but you will probably want to update the others.

3. Open your `terraform.tfvars` file for editing. As we've seen previously, this file will be used to set the variables that will be used in the various TF files.

4. Before I talk about some of the optional variables you may want to add, let's start with the ones that are already in the file. `domains_to_allow` will limit permissions that are granted to users in this list of domains. I'm only configuring my `gcp.how` domain, so that's my only value.

5. `billing` and `audit_data_users`: If you used the recommended names for these two security groups, all you should have to do is change the domain names.

6. Fill in your `org_id` and `billing_account` values. Remember, they can be retrieved using the following commands:

    ```
    gcloud organizations list
    gcloud beta billing accounts list
    ```

7. `terraform_service_account`: Check your notes. This is a value that you recorded when you performed the bootstrap project setup. It was one of the outputs in that step.

8. `default_region`: This is your preferred default region for the common projects. Make sure that the region supports BigQuery since we are syncing some data to it.

9. `scc_notification_name`: If needed, set a unique name for the SCC notification being created. Notification names need to be unique.

    That's it for the required variables. Now, let's look at some you may want to consider. For a full description of all the variables and their use, see the `variables.tf` file. Let's look at a few key options.

10. `parent_folder`: If you aren't applying this to the root of your organization, provide the ID of the parent folder where it should be applied.

11. `data_access_logs_enabled` (a default value of `true`): This enables Data Access logs on all GCP services. That means that any time a user modifies data, a log is recorded. Data Access logs in GCP are typically disabled by default, and although enabling them provides a lot of useful information, the extra logs are a billable expense. You can decide if you would like to disable this or not.

12. `Enable_hub_and_spoke`: We will talk more about hub and spoke network designs later. I'm not going to use it in my example, so I will leave this set to the default of `false`.

13. `create_access_context_manager_access_policy`: By default, the example configuration creates and enables an Access Context Manager to help restrict access to our environment. In this case, it will limit what user rights are granted to accounts that belong to your domain. If you want to disable this, set this value to `false`. We will discuss Access Context Managers later in this book, but if you're like more information, go to `https://cloud.google.com/access-context-manager/docs/overview`.

Excellent! With the variables all set, it's time to finish the `1-org` segment of the Example Foundation.

14. Commit the changes you've made. Make sure you are still in the `gcp-org` folder, then add the files to your local Git `plan` branch and commit them:

```
git add .
git commit -m 'Initial org commit'
```

15. Push your plan branch to the CI/CD project repository, which will trigger a plan for your environment:

```
git push --set-upstream origin plan
```

16. Check your notes for the bootstrap build and copy the name of your CI/CD project. Then, use it in the following URL:

`https://console.cloud.google.com/cloud-build/builds?project=YOUR_CLOUD_BUILD_PROJECT_ID`.

17. This is a list of the Cloud Build jobs. Locate the build job that corresponds to the `gcp-org` source that has an appropriate created time. Click on the job to view its details.

18. Scroll through the build log and examine the plan. If it looks good, apply the change. If you see errors related to the validator, ignore them at this point.

19. To apply the changes, you create a production branch and push it. This will fire the apply trigger and apply the changes to production, which includes the common folder:

```
git checkout -b production
git push origin production
```

20. Wait a minute and revisit your build jobs: `https://console.cloud.google.com/cloud-build/builds?project=YOUR_CLOUD_BUILD_PROJECT_ID`.

21. Wait for the latest production build job to complete, which will likely take more than 5 minutes but less than 10, then click to investigate the logs.

Fantastic! At this point, you should have the shared folder and all its related projects in place. To see them, go to **Navigation menu | IAM & Admin | Manage Resources**, and investigate the new structure.

Now that you know how all this stuff works, I'm going to let you build out the prod/non-prod/dev environments using the steps at `https://github.com/terraform-google-modules/terraform-example-foundation/tree/master/2-environments`. You'll see that the process is almost identical to what you just did:

1. Check out your `gcp-environments` repository.
2. Create a plan branch.
3. Copy in Google's starter files, Cloud Build YAML files, and wrapper bash script.
4. Edit a few TF variables.
5. Commit the changes to your `plan` branch.
6. Push the changes to the remote origin repository.
7. If you're happy with the plan, use the same files to create a development branch and push it to implement it.
8. Lather, rinse, and repeat with the non-production branch and environment.
9. Lather, rinse, and repeat with the production branch and environment.

Before you tackle these steps, I want to point out a few interesting things and variable settings. First, you'll notice that this time, there are a couple of subfolders in `gcp-environments`:

- envs: These are our root modules. Each one matches one of our three environments and contains a fairly basic `main.tf` file that essentially says, "*apply the env_baseline module with these names and tags.*"

- `modules/env_baseline`: This is what's doing most of the work. It creates a folder for the environment and uses the project factory to create a monitoring project, a couple of shared VPC projects, and a secret project. These are the same common, environment-specific projects we discussed earlier.

If you examine the variables file in the top-most `gcp-environments` folder, you will find some of the same configurations we've changed in other projects (`org_id`, `billing_account`, and so on).

Before you add and push your plan branch, make sure that you update the root `terraform.tfvars` file. You will notice that one of the variables references a security group that doesn't exist yet, designed to hold users that need to be monitored at the environment level. We will discuss this group later, but for now, add the group with no specific permissions to your Google Cloud environment.

Also, the `env_baseline` TF files, once again, apply labels to the projects. You may want to edit those before implementing the plan.

I'm going to hold off on implementing *3-networks* and *4-projects* as we need to discuss a few other things first.

Fabulous work! You now have a CI/CD process in place to manage infrastructure in a toil-free, automated way. You can create TF scripts to set up Google Cloud resources and you've completed the fifth step in our 10-step foundation plan. Hey, you're halfway there!

Summary

In this chapter, we made fantastic progress laying our Google Cloud foundation by completing step 5 and creating a resource hierarchy to control logical organization, from Google's 10-step recipe. Not only did we learn how to plan and build a resource hierarchy, but we also simplified all our future steps by learning how to automate infrastructure management using a popular IaC product: Terraform. To make using TF easy and standard, we also implemented a CI/CD pipeline to allow us to continue working with our infrastructure from a central pair of projects.

If you want to keep moving through the checklist steps with me, your tutor, please move on to the next chapter, where you will learn how to use GCP security roles to control access across your presence in Google Cloud.

5
Controlling Access with IAM Roles

Do you remember in the last chapter when you got to figure out where you and all your fellow employees would sit in the new office building? Remember how we decided that a good way to attack the problem would be to examine how employees should be grouped both from a logical team sort of perspective and with an eye on security needs? Well, just like when building our resource hierarchy, Google Cloud is going to require the same type of thinking when we start to lay in our access control.

Google believes that the way you get security right is by thinking in cybersecurity terms from the ground up, and from day zero forward. As a result, Google Cloud has created an industry-leading, zero-trust architecture with a proven track record of security and reliability. Let's do it.

In this chapter, we're going to start with an introduction to **Identity and Access Management (IAM)**, and then continue to use Terraform to ensure security is applied where and how we need it, and thus stratify the sixth step in Google's foundation-laying plan:

- Understanding IAM in Google Cloud
- Step 6 – Adding IAM trust boundaries to the resource hierarchy

Understanding IAM in Google Cloud

Imagine you are taking a trip, maybe heading on vacation to Walt Disney World in Florida. You head to your local airport with your ID and your plane ticket in hand, and you make your way to security. Typically, the first officer you talk to at security performs two major tasks. First, they check your ID to make sure you are who you say you are. Once you are authenticated, they then check your ID against your ticket and determine whether you are authorized to proceed through security.

Even if you are passed through security, your authorization level will only allow you to go into specific airport areas and do specific things. You likely can't carry a gun with you through security because you don't have that level of access, just like there will be parts of the airport you aren't allowed into.

Now, let's transition away from airports and back into Google Cloud. Back in foundational step one, we configured our Google Cloud identity management. Then, in step two, we added an initial set of users and security groups. What we didn't address past a few key groups was how Google Cloud handles authorizing users and groups to do things in Google Cloud. Google Cloud IAM is all about who can do what and where – something like the following graphic.

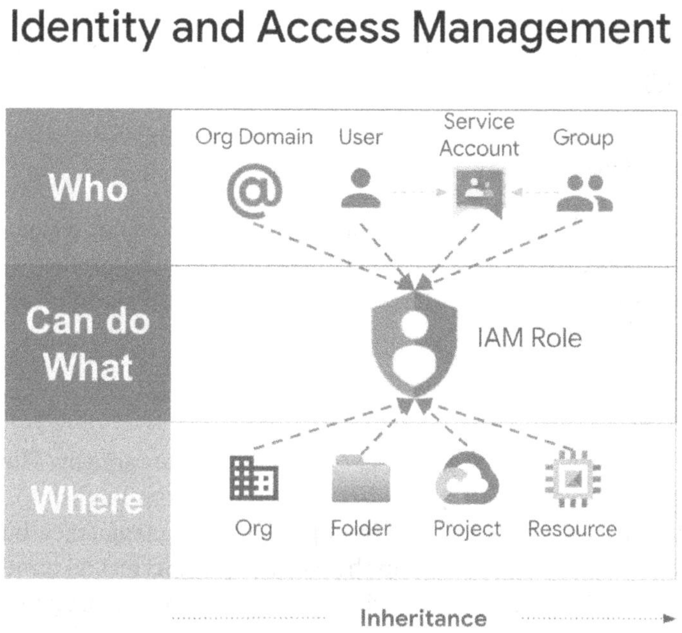

Identity and Access Management

Figure 5.1 – Who can do what where

Let's start with the who.

Who?

The *who?* in authentication and authorization in Google Cloud is technically defined by a **principal**. Google Cloud supports several different principal types, including the following:

- **Google Account**: Essentially a valid user in Google Cloud. Typically, these users come from a Google Workspace, Cloud Identity domain, or a Gmail account.

- **Service account (SA)**: An account managed in Google Cloud used to authenticate applications or compute workloads into Google Cloud services.

- **Google group**: A collection of Google accounts and service accounts

- **Google Workspace account**: A collection of Google accounts defined as part of a Google Workspace.

- **Cloud Identity domain**: A collection of Google accounts defined as part of a Cloud Identity domain, much like a Google Workspace account.

- **All authenticated users**: `allAuthenticatedUsers` is a special placeholder representing all service accounts and Google accounts, in any organization (not just yours), including Gmail.

- **All users**: `allUsers` is a special placeholder representing all authenticated and unauthenticated users (anonymous access).

Most principals are created and managed in the identity management system we configured in step one. Exceptions include Gmail accounts, the special `allAuthenticatedUsers` and `allUsers` placeholders, and Google Cloud service accounts.

Service accounts

I was teaching the other day when the topic of Google Translate came up. How hard would it be to create a piece of code that could translate text from English to Spanish? The group I was training worked with many different programming languages, but JavaScript seemed to be a lingua franca. So, I pulled up the Node.js (JavaScript) documentation for the Google Cloud Translate library, and right on its home page (`https://cloud.google.com/nodejs/docs/reference/translate/latest`) is a quick example. I copied the code, put it in a code editor on my laptop, changed the `ru` Russian language code to the `es` Spanish code I needed, and ran the demo on my laptop.

The demo worked perfectly, but how did code running on my laptop access a service provided by Google Cloud? The answer: service accounts.

Services accounts are used to provide non-human access to Google Cloud services. Unlike user Google accounts, they don't belong to your Google Workspace or Cloud Identity domain and are created and managed inside of Google Cloud projects. Also, they have no associated password and cannot log in via a browser. Authentication is accomplished via public/private key pairs.

There are three service account types:

- **User-managed service accounts**: Created by you using the console, CLI, or API. By default, any project can create up to 100 SAs, though that quota may be raised. You can choose the name of your SA, which will be combined with `project-id` to create an SA email address username in the format `name-you-choose@project-id.iam.gserviceaccount.com`.

- **Default service accounts**: Created by Google to run App Engine (`project-id@appspot.gserviceaccount.com`) and/or Compute Engine (`project-number-compute@developer.gserviceaccount.com`). As a best practice you likely don't want to use these accounts, but you shouldn't try and delete them – it will break stuff.

- **Google managed service accounts (robots)**: Created by Google in conjunction with various Google Cloud services and used when those services act on your behalf. Examples would include robot accounts set up behind Kubernetes and Compute Engine, Cloud Storage, or Data Fusion.

So, the *who?* In Google Cloud, the principal, is usually a person or a service account. With that established, let's talk about *what* the principal can do.

Can do what?

In my English-to-Spanish translation example, the SA my code ran under needed to access the translation service through a particular project, but it didn't need to do much more than that. From a principle of least privilege perspective, I'd like to be able to grant that specific level of access at a fine-grained level. Enter IAM.

Google Cloud IAM allows you to granularly grant different levels of access to specific resources. If a user or SA needs access to a resource, you first need to determine what level of access they require. **Permissions** define precisely what a principal can do, and take the form `service.resource.verb`. Then, instead of allowing the direct assignation of permissions, Google combines permissions into **roles**. So, you don't grant a user a permission, instead you create an IAM **policy** and bind the user to the role, and the role defines a set of permissions. Typically, rather than configuring IAM policies for individual users, you will gather your users into security groups and then configure the binding policy for the group, rather than an individual.

In the grand scheme of things, this isn't new or revolutionary. Controlling security by granting access via role bindings is used in a lot of security-related systems. One thing a little different here is that the user base and authentication mechanism are defined outside of Google Cloud, but the roles, role bindings, and authorization are defined inside. Regardless, *can do what?* is something that's going to take thought and consideration. We will get more into specific role recommendations later in this chapter. For now, let's dig a little deeper into how Google Cloud roles work.

Google defines three role types: **basic**, **predefined**, and **custom**. A great reference related to this entire Google Cloud role discussion can be found here: `https://cloud.google.com/iam/docs/understanding-roles`.

Basic roles, sometimes called primitive roles, were the original Google Cloud roles, and they predate today's IAM system. Their strength, and their weakness, is that they are very coarse-grained. They are typically granted at the project level where they apply to most, if not all, of the services in the project. The three basic roles are as follows:

- **roles/viewer (Viewer)**: Read-only access to most project resources, data, and configurations.

- **roles/editor (Editor)**: All viewer permissions, plus the ability to modify most resources, data, and configurations. In addition, editors can create and delete resources.

- **roles/owner (Owner)**: All editor permissions, plus the power to manage roles and permissions for others, and to configure project billing.

The problem with basic roles is that they are too coarse-grained. How many times have I heard the words, *"Alicia couldn't read that contract file in Cloud Storage but no worries, I made her a viewer and she can now."*?

Palm to forehead.

From a laziness perspective, basic roles are very attractive. Yes, making Alicia a viewer allowed her to see that file, and it was an easy fix, but Alicia can now also read the tables in that Spanner database, see all the settings for all the project Compute Engine VMs, and run queries in BigQuery. Did she need all those extra permissions? Probably not. One thing I know is that making Alicia a viewer likely violated the principle of least permissions.

There are a few cases where users doing experimentation or development might use things such as viewer or editor, but even in those cases, the basic roles make me nervous. A much better option tends to be to grant permissions by putting people into groups and then binding the groups to a set of predefined roles.

Where basic roles apply to everything in the project, **predefined roles** are fine-grained roles tied to individual Google Cloud services, and to specific jobs related to those services. Predefined roles are created and managed by Google, so as cloud features evolve and change, Google can update and adapt the roles. Did you take a peek at the page I recommended earlier in this section? If not, do so now: `https://cloud.google.com/iam/docs/understanding-roles`. Did you notice how long that page took to load? That's because there are a lot of Google Cloud services, and each one comes with its own set of predefined roles.

You remember Alicia and her need to access files in Cloud Storage? On the right side of the preceding page is a list of predefined roles grouped by service type. Scroll through the list and click on **Cloud Storage roles**. In addition to this huge list of roles all on a single page, many Google Cloud products have documentation pages dedicated to access, roles, and permissions for that single product. If the huge list doesn't provide enough information for you to make a decision, do a little Googling. For example, I just did a search for `GCP Cloud Storage IAM` and the first result is the Cloud Storage detail page here: `https://cloud.google.com/storage/docs/access-control/iam-roles`.

Note – When Using Google Search to Look for Google Cloud Related Information, Prefix Your Search with GCP

Google Search knows that **GCP** is short for **Google Cloud Platform**. Searching for `Compute` returns a whole mess of non-cloud results, starting with a definition. Searching for `GCP Compute` gets you right to the Compute Engine documentation.

Looking in the Cloud Storage roles on either of the preceding pages you will see that Cloud Storage calls a file an object. Essentially, a Cloud Storage object is the file plus any metadata related to the file. Looking through the object roles, it won't take you long to locate `roles/storage.objectViewer` (Storage Object Viewer), which grants access to view objects and their metadata. The column to the right of the role name and description details the specific individual permissions the role grants – in this case, the following:

- `resourcemanager.projects.get`
- `resourcemanager.projects.list`
- `storage.objects.get`
- `storage.objects.list`

With that, Alicia could see the project and if she had the URL to the bucket, she could see (`list`) the files and read (`get`) any she needed to view. Interestingly, she couldn't get a list of the project's buckets, so that's something to keep in mind.

Best Practice – Use Predefined Roles Where Possible

Google recommends as a best practice that predefined roles be used wherever possible. Google has put a lot of thought into how resources and roles relate and set up the predefined roles accordingly. Also, as the cloud evolves, Google continues to update and maintain the roles, making forward-moving role management easier.

If predefined roles won't work for some reason, such as one doesn't grant quite enough permissions and the next grants too many, then consider creating your own custom roles.

Custom roles are created and maintained by organizations rather than Google and they contain user-specified sets of individual permissions. I see custom roles used a lot by organizations, but all too often they are used for tradition or ignorance reasons, more than for any compelling need.

"We've always used custom roles."

"Listen, I looked at that huge list of predefined permissions, and it just looked too hard, so I went custom instead."

Likely, you will regret an over-dependence on custom-created roles. Not only are they difficult to construct and get right in the first place, but then you also have to manage them over time and through cloud evolutionary changes.

We could have used a custom role when solving Alicia's file access problem. Perhaps you don't like that as a Storage Object Viewer she can see the files in the bucket, but can't get a list of available buckets from the project through the UI. One possible solution would be to create a custom role very close to Storage Object Viewer. You can't edit predefined roles, but you can clone them and use the clone as a foundation for new roles. You could open the Storage Object Viewer role, make a copy, and then add in the permission to list buckets, then bind Alicia to your new custom role and she will be able to list buckets and files and read the files themselves.

Excellent – now that we have a little foundational knowledge on IAM and role usage, let's use what we've learned to secure the resource hierarchy we've constructed.

Step 6 – Adding IAM trust boundaries to the resource hierarchy

It's time to add some access control into our resource hierarchy in Google Cloud. The details of how and where precisely you will apply permissions in the access management stage of your Google Cloud setup will vary based on your organizational needs. After the last section, you probably have at least some ideas about how you will use the various role types, but I'm sure there are a lot of gaps that need filling.

You may not have realized it but our Google Cloud security posture started much earlier than this chapter. In our first major Google Cloud foundation-laying step, we set up our user identities, linked in an authentication service, and created a Google Cloud organization, all so we could have a way to control user management and authentication. From there, we set up some admin users and groups, established administrative access to Google Cloud, and then in the last chapter, we proposed and built a resource hierarchy, based on Google's security foundations guide. The hierarchical plan we came up with looked like the following figure:

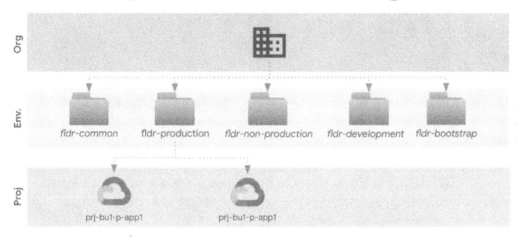

Figure 5.2 – A possible resource hierarchy

You might have tweaked the design, perhaps adding a layer of business unit folders above the environments, but as long as you kept both security and organization in mind, the design will do just fine. What's nice about this design is both its simplicity, with one folder for each of the major environment types, a folder for the CI/CD projects, and a folder for all our common shared projects, and its flexibility in letting us assign major permissions and other policies at a high level, so we can get security in place sooner rather than later. If you used the example best practice scripts from Google's Example Foundation, then you have already built some security into what you've constructed. We will talk about that, and more as we see to our access control in Google Cloud. Let's start by examining security roles and what they are really telling us.

Reading a security role

If you recall our earlier example, where Alicia needed access to a file in Cloud Storage, we solved the problem by making her a Storage Object Viewer. If you look at the big page of predefined Google Cloud security roles (`https://cloud.google.com/iam/docs/understanding-roles#predefined`), then the entry for Storage Object Viewer resembles the following:

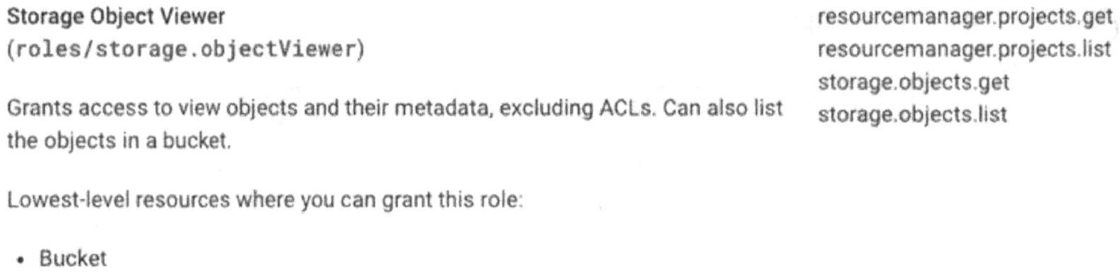

Storage Object Viewer (roles/storage.objectViewer) Grants access to view objects and their metadata, excluding ACLs. Can also list the objects in a bucket. Lowest-level resources where you can grant this role: • Bucket	resourcemanager.projects.get resourcemanager.projects.list storage.objects.get storage.objects.list

Figure 5.3 – Storage Object Viewer

In the upper left, we see the human-readable name, **Storage Object Viewer**, and the technical name, `roles/storage.objectViewer`. The individual permissions, as discussed earlier in this chapter, are listed down the right side, and in the lower-left corner, we see the *lowest-level resources* where the permission may be applied.

The lowest-level resource identifies where, in the context of the resource hierarchy, this permission may be used. Essentially, it's saying that I could grant Alicia the Storage Object Viewer permission on a single bucket, at the lowest level, or I could grant her the permission on the project so she could see all the files in all the project buckets. Lastly, I could go all the way up and grant her the permission at the organization level, so she could read any file in any bucket in any project in any folder across the entire organization.

The supplied technical name is useful because if we need to apply this permission with the command line, code, or with Terraform, it lets us know the value to use. If you examine the permissions setting with Terraform for a bucket you'll see: `https://registry.terraform.io/providers/hashicorp/google/latest/docs/resources/storage_bucket_iam#google_storage_bucket_iam_member`. So, if we wanted to directly grant Alicia the Object Viewer role, we could do it with something like this:

```
resource "google_storage_bucket_iam_member" "alicia_r" {
  bucket = "some-bucket-name"
  role = "roles/storage.objectViewer"
  member = "user:alicia@gcp.how"
}
```

Though we'd be much more likely to drop her into a group and control her permissions from there. Speaking of groups…

Use groups where you can

We said in an earlier chapter that one way to simplify your user management was to place users in groups and then bind the roles to set the permissions at the group level. It's also possible in GCP to nest groups in other groups, so when building out your security role assignments, that's a nice trick to make group creation more modular.

Before we get into setting the access control for our hierarchy, let's establish some naming guidelines.

If you open the naming standards document we created back in *Chapter 2, IAM, Users, Groups, and Admin Access*, the last couple of entries you should see relate to standard folder and project names. Let's add a few new entries for groups, custom roles, and service accounts:

Resource	Naming convention
Group	`grp-gcp-group-label` Example: `grp-gcp-billing-admin`
Role	`rl-function-role-label` Example: `rl-limited-compute-admin`
Service Account	`sa-env-code-business-code-details` Example: `sa-p-dep4-cool-app-bkt-reader`

Table 5.1 – Group, role, and service account names

You'll notice in my recommended name for groups that I've added a gcp to the prefix. I'm doing that to keep the groups that are specific to Google Cloud separate from the groups that might be related to other parts of the organization. Remember, these groups will be created in your user store and authenticated with your identity provider. You'll likely have groups that are broad-reaching across your business, with importance both in GCP and in other areas of the organization. Those groups might have names such as sales-members or accounting-admins. As such, it's handy if you can easily identify the groups you are creating specifically for use in Google Cloud, thus the prefix.

If Alicia is one of several staff members who need to read contracts across one-to-many buckets, then it would be a good idea to create a group to help them do just that, perhaps `grp-gcp-contracts-team`. To create the group itself, you'd need to determine the best approach given your environment. Your organization likely already has policies and procedures around security group creation, in which case you'll need to do some research there. Perhaps you submit the request to a security team, and they create the group and assign the membership in Active Directory. If you are managing your groups in Cloud Identity, then you can use the web UI to create groups over at `https://admin.google.com/ac/groups`. Google does have a group creation helper Terraform script for use with Cloud Identity over with the other Terraform blueprints (`https://cloud.google.com/docs/terraform/blueprints/terraform-blueprints`) and it can get the job done, but make sure to pay attention to its limitations.

Just like creating the group itself, you should check with your org on how user-to-group assignments are normally handled. Typically, someone on the security management side of the org would again use the user/group management identity service directly to accomplish the task, so they would assign the users to the group in AD or Cloud Identity.

Once the group has been configured and the personnel associated, we need to get the role assignments handled over in Google cloud. Groups typically will receive a set of role assignments, to allow the folk in the group to do their jobs while following a principle of least permissions model. This can most easily be accomplished with Terraform, though using the IAM page in the Google Cloud Console or `gcloud identity groups memberships` will also work.

For help setting up group role bindings with Terraform, look in Google's Terraform Blueprints, specifically in the IAM sub-section. There, you will see that Google has scripts to help set permissions on a set of storage buckets, for projects, for folders, and for the whole organization. To take our `grp-gcp-contracts-team` group and grant it read access on a set of buckets, we could use the Storage Buckets IAM Terraform script:

```
module "storage_bucket-iam-bindings" {
  source          = "…details omitted for space"
  storage_buckets = ["bkt_1", "bkt_2"]
  mode            = "additive"

  bindings = {
    "roles/storage.objectViewer" = [
      "group: grp-gcp-contracts-team@gcp.how"
    ]
```

```
    }
}
```

Notice that both the `bindings` and `storage_buckets` sections can accept arrays of values. An especially nice feature of the Google blueprint IAM Terraform scripts is the `mode` option. When the mode is `additive` as in this example, it adds the specified permissions to whatever permissions currently exist on the buckets. If the mode is `authoritative`, then the script's application overwrites any already assigned permissions with the specified authoritative set.

Good – with an idea of how we could leverage Terraform to apply GCP security roles, let's talk about some major security groups we might want to create.

Google Cloud starter security group ideas

You might recall that we created some security groups back when first establishing our Google Cloud foundational administrative access, including the following:

- `gcp-billing-admins`: Full control over setup, configuring, and monitoring billing accounts.

- `gcp-developers`: Can design, code, and test applications.

- `gcp-devops`: Create and manage CI/CD pipelines and manage infrastructure for applications.

- `gcp-network-admins`: Create and manage networks, subnets, firewalls, and other network-related resources. We're going to need this in our next chapter.

- `gcp-organization-admins`: Google Cloud organization administrators, with full control over the logical structure of your organization in GCP.

- `gcp-security-admins`: Core security personnel, responsible for setting security policies across the org structure.

If you read through the Google Cloud security foundation's guide, the bases of the Example Foundation Terraform scripts we used in the last chapter, you will see that they suggest several additional security groups, including the following:

- `grp-gcp-scc-admins`: Administrators for the Security Command Center.

  Security Center Admin Editor on the SCC project.

- `grp-gcp-monitoring-admins`: Have the ability to use Cloud Monitoring across the various monitoring projects. (Note: the current version of the Google Example Foundation projects named this group `gcp-monitoring-admins`.)

Monitoring Editor on the production, non-production, and development monitoring projects.

- `grp-gcp-security-reviewers`: Security team members responsible for reviewing org-wide security.

  Security Reviewer on the whole org.

- `grp-gcp-billing-viewers`: Mostly for people on the finance team who need access to view spend, billing alerts, reports, and the like.

  Billing Account Viewer on the organization.

  BigQuery Data Viewer / BigQuery User on the billing project.

- `grp-gcp-network-viewers`: Network team members who need to review network configurations.

  Compute Network Viewer on the root organization node.

- `grp-gcp-audit-viewers`: Security and audit team members who need to see audit logs.

  Logs Viewer / Private Logs Viewer / BigQuery Data Viewer on the logging projects.

- `grp-gcp-platform-viewers`: Anyone who needs the ability to view resources across the entire organization.

  Viewer on the root organization node.

Instead of, or in addition to, that global developers group created when we established initial admin access, we might want to add other developers groups, something like `grp-gcp-`**business-code-environment-code**`-developers`. Here, we could put developers who need access across a portion of the projects in a given environment. The environment-specific permissions would then be set based on need.

Terraforming the permissions

I'm not going to walk through Terraforming all these proposed groups and permissions, but let's do a couple to see how it would work with the CI/CD process we have in place. After you know the approach, you can plan your groups, permissions, and use Terraform to put things together.

For my examples, I'm going to set up the billing viewers and security reviewer's groups and permissions.

First, you need to construct the groups themselves. To do that, you'd go to your source of truth for users and create the groups. For me, that's Cloud Identity. I head over to `https://admin.google.com/`, locate the **Directory | Groups** page, and create the two groups, something like the following:

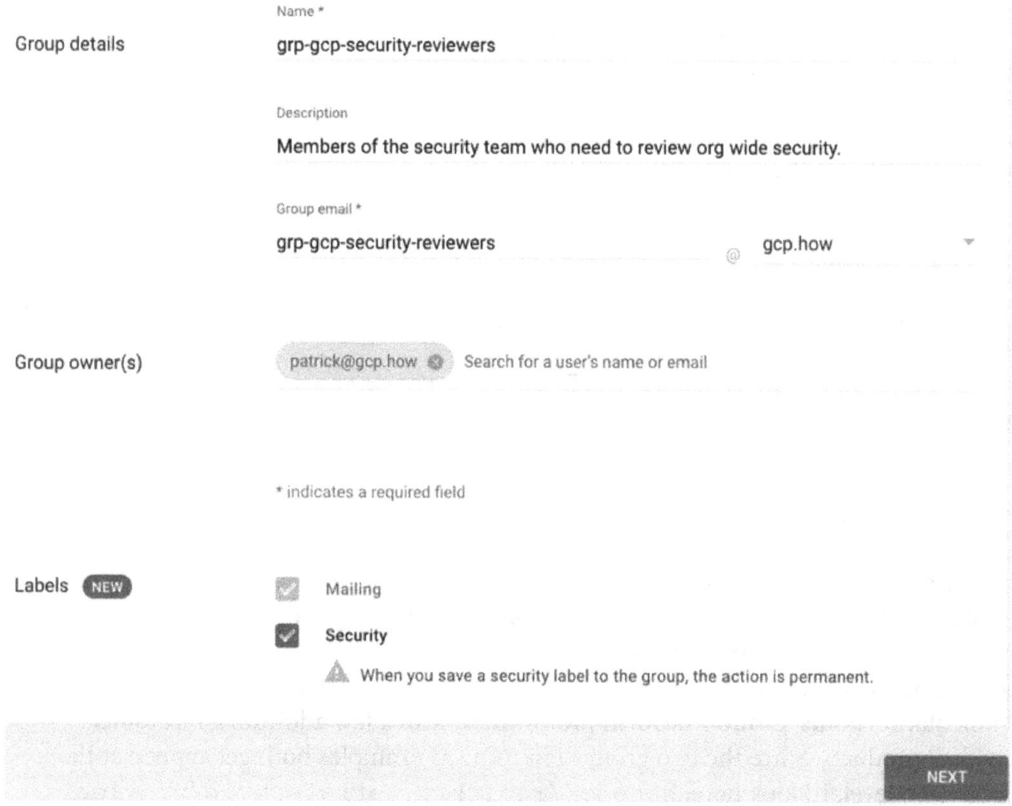

| Group details | Name *
grp-gcp-security-reviewers |
| | Description
Members of the security team who need to review org wide security. |
| | Group email *
grp-gcp-security-reviewers   @   gcp.how |
| Group owner(s) | patrick@gcp.how ⊗   Search for a user's name or email |
| | * indicates a required field |
| Labels NEW | ☑ Mailing
☑ Security
⚠ When you save a security label to the group, the action is permanent. |
| | NEXT |

Figure 5.4 – New security reviewers' group

With the groups created, remember how the Terraform scripts and our CI/CD project work. Start by deciding where each group needs to be applied and what exact roles it will need. Is this something that's org-wide, or something that's tied to the three development environments? Remember, at this point, we have a `gcp-org` repo with scripts that impact the entire organization and the common folder, and we have a `gcp-environments` repo for scripts impacting the dev, non-prod, and prod environments. In this case, these are org-wide security settings and groups, so the process would be essentially this:

1. Clone the `gcp-org` repo.
2. Check out the `plan` branch.
3. Make the requisite changes to the Terraform scripts.

4. Add the updated files to the branch and commit the changes.

5. Push the `plan` branch to the origin to generate a Terraform plan.

6. Verify the plan generated by the trigger.

7. If all's good, check out the `production` branch, merge the `plan` branch into it, and push it to the origin to apply the change.

8. Verify the plan's application and that there were no errors.

So, I start by opening my terminal window and cloning down the `gcp-org` repo from my CI/CD project, and checking out the `plan` branch:

1. Clone down the `gcp-org` repo using the terminal. If you already have a copy, you might consider removing it, if you're not sure of its status:

    ```
    gcloud source repos clone gcp-org \
    --project=<CICD_PROJECT_ID>
    ```

2. Check out the `plan` branch:

    ```
    cd gcp-org
    git checkout plan
    ```

With the `plan` branch down and ready for editing, I open my editor and have it open the `gcp-org` folder. Under that, in the `envs/shared` folder, I already have an `iam.tf` file containing the IAM Terraform configurations. These configurations rely on Terraform's Google Cloud IAM resources, found for the most part in the Terraform for Google documentation under **Cloud Platform | Resources**, with a few additional bits under individual products. Since the two groups I'm using as examples both get applied at the organization level, I'd look here: `https://registry.terraform.io/providers/ hashicorp/google/latest/docs/resources/google_organization_iam`.

To make things easy in this case, it turns out the example `iam.tf` from Google already has policies for the groups I've selected, and all the groups in my list. If you are not using the example scripts, you can still look at them to see how they take care of the assignments. If you scroll through the file, you'll see there's a section that starts with a `Billing BigQuery - IAM` header comment that looks like this:

```
resource "google_project_iam_member" "billing_bq_user" {
  project = module.org_billing_logs.project_id
  role    = "roles/bigquery.user"
  member  = "group:${var.billing_data_users}"
}
```

```
resource "google_project_iam_member" "billing_bq_viewer" {
  project = module.org_billing_logs.project_id
  role    = "roles/bigquery.dataViewer"
  member  = "group:${var.billing_data_users}"
}
resource "google_organization_iam_member" "billing_viewer" {
  org_id = var.org_id
  role   = "roles/billing.viewer"
  member = "group:${var.billing_data_users}"
}
```

That's a nice example of how you use the `google_organization` and `project_iam_member` resources, and how they bind the roles to the billing data user's group. Later in the same file, you will find another section doing something similar for the security reviewer. You'll notice that in both sections the only requirement is to drop in values for the `gcp_security_reviewer` and `billing_data_users` variables.

To make the changes to the variables file, take these steps:

1. Open the `envs\shared\terraform.tfvars` file. You'll notice that you already have a `billing_data_users` variable set. You'll need to update that variable, changing it to your new group, and you'll need to add the new `gcp_security_reviewer` variable. Your domain name will be different than mine:

   ```
   gcp_security_reviewer =
   "grp-gcp-security-reviewers@gcp.how"
   billing_data_users = "grp-gcp-billing-viewers@gcp.how"
   ```

2. Add the updated files and commit the change:

   ```
   git add .
   git commit -m "IAM for billing view and sec review"
   ```

3. Push the update to the CI/CD project and generate the plan:

   ```
   git push -u origin plan
   ```

4. To view and approve the plan, visit the Cloud Build history page in your CI/CD project: `https://console.cloud.google.com/cloud-build/builds?project=<YOUR_CICD_PROJECT_ID>`.

5. In the plan, you'll see how the updates are being applied, first by removing the old billing viewer permission and recreating it with the new. Then there is also a new binding for the security reviewer. If all looks good, merge the plan into production and push the change:

```
git checkout production
git merge plan
git push -u origin production
```

6. To verify the successful run of the apply trigger, check the Cloud Build history: `https://console.cloud.google.com/cloud-build/builds?project=<YOUR_CICD_PROJECT_ID>`.

That's it, we've successfully added a pair of new security groups, and set their IAM permissions where and how we need with Terraform. Nice job. If you want to see the results, you can open and examine the IAM settings in your organization and for the `prj-c-billing-logs` projects.

But what if simply applying a role binding won't exactly solve your problem and you really need to do some fine-tuning? Well, you might be able to solve your problem with conditions.

Fine-tuning IAM permissions with conditions

Though most of our IAM configurations will be set at the group and role levels, there are some other things we can do to help refine user permissioning. Org policies are one tool but I'm going to talk about them in a later security chapter. For now, let's talk IAM conditions.

IAM conditions allow us to fine-tune permissions by implementing a level of attribute-based access control. Essentially, you add a condition using something called the **Common Expression Language** (**CEL**), which expresses a logical test and, based on the test results, the IAM role will be selectively applied. Common types of conditions would be tied to things such as resource types, names, tags, and date/time of the request. Details of the CEL spec can be found here: `https://github.com/google/cel-spec/blob/master/doc/langdef.md`.

Let's look at some example conditions, and then how to use the conditions in Terraform.

Do you remember Alicia? We granted her and her contract readers group access to a series of buckets by explicitly naming them. What if we wanted instead to grant her access to any bucket with a name ending in -contracts? Well, if that was the need, then we could use a condition to grant the permission:

```
resource.type == "storage.googleapis.com/Bucket" &&
resource.name.endsWith("-contracts")
```

Before I provide another example, does that last example make you nervous? It does me. Attribute-based permissions are a double-edged sword. They can solve some permission issues and provide detailed refinements, but I'm not a fan of a condition that could inadvertently grant someone access they shouldn't have. In the preceding case, if someone in some part of the org happens to create a bucket with that naming suffix, Alicia and her group will all get access. The issue is somewhat mitigated by the fact that we are funneling resource creation through Terraform, and will likely be able to control naming from there, but still. If you are granting permissions on things such as resource names, make sure you also properly control the naming.

Another example would be a restriction on access to a 9-5 New York window of working hours:

```
request.time.getHours("America/New_York") >= 9 &&
request.time.getHours("America/New_York ") <= 17
```

As a final example, what if you wanted to grant the contract reader team access to a particular bucket but you wanted the access to expire at the end of 2025. Here, the condition might look like this:

```
request.time < timestamp("2025-01-01T00:00:00Z")
```

Okay, now that we've seen some conditions, how would they work with Terraform? That part is actually easy. All of the Terraform direct resources and those Google has extended in their Google Cloud helpers will accept conditions. For a Terraform native example, check out https://registry.terraform.io/providers/hashicorp/ google/latest/docs/resources/google_organization_iam#google_ organization_iam_binding. If we were using the condition with our Google Terraform blueprint example, then we would change bindings to conditional_ bindings, like this:

```
module "storage_bucket-iam-bindings" {
  source               = "...details omitted for space"
  storage_buckets = ["bkt_1", "bkt_2"]
```

```
mode            = "additive"

conditional_bindings = [
  {
    role = "roles/storage.objectViewer"
    title = "expires_2025_01_01"
    expression =
  "request.time < timestamp('2025-01-01T00:00:00Z')"
    members = ["group: grp-gcp-contracts-team@gcp.how"]
  }
] }
```

Not bad. Conditional permissions offer a nice way to tune permissions based on attributes, but they do have some limits. For a start, not all resources support conditional role bindings so make sure to check the supported resource list before trying to use this feature: `https://cloud.google.com/iam/docs/conditions-overview#resources`. Also, there are limits to how many logical operators a given condition can have (12), and how many variations of the same member-to-role binding are allowed (20). Check out the latest quotas and limits for IAM here: `https://cloud.google.com/iam/quotas`.

> **Note – Don't Accidentally Empower the User to Remove the Condition**
>
> When using IAM bindings with a condition, keep in mind that resource roles containing the `setIamPolicy` permission have the power to modify IAM bindings for that resource and as such, users with that permission could simply remove any conditional restrictions placed on them.

Before we leave the conditional IAM permissions discussion, let's take a few minutes to discuss tagging.

Linking conditional IAM permissions to resource hierarchy tags

A final way of using conditions deals with something we've never discussed: **resource tags**. On the surface, tagging looks a lot like something we have seen, labeling. Both tagging and labeling leverage *key-value* pairs to help identify resources, but while labels can be attached to most resources and tend to be used for better log searchability and finer-grained billing export analysis, few resources currently support tagging (`https://cloud.google.com/resource-manager/docs/tags/tags-supported-services#supported_service_resources`) and it is designed for use with IAM conditional permissions and organizational policies.

> **Resource Tagging Is Different than Network Tagging**
>
> Unfortunately, **tagging** isn't a unique term, even within Google Cloud. For years in GCP, network firewalls have used a form of tagging to grant or deny access to Compute Engine VMs. It is a totally different, similarly named, concept. Tagging in relation to the resource hierarchy, as it's discussed in this section and as it is used with organizational policies, is something new in Google Cloud, so don't confuse your types of tagging.

Since both IAM permissions and org policies can be linked to tags, their creation should be tightly controlled, and automated. For tagging related permissions, you can find Tag Administrator, User, and Viewer in the Resource Manager section of the big roles page: https://cloud.google.com/iam/docs/understanding-roles#resource-manager-roles.

To use a tag, first, it and its values need to be defined at the organizational level. Make sure to check the documentation on what characters are allowed in both the names and values: https://cloud.google.com/resource-manager/docs/tags/tags-creating-and-managing#creating_tag.

Creating tags can be accomplished in one of three ways: using the **IAM | Tags** page in the console, using the Google Cloud gcloud resource-manager tags command, or via the API, which is how our good friend Terraform gets tags done.

When you create a tag, you specify a short name, a description, and an organizational ID. Google will create the tag and assign it a globally unique (and never reused) name, and a more human-readable namespaced name.

Creating a value for a tag will again need a short name (with slightly different acceptable character requirements) and a description, and it will again get a unique name and a more readable namespaced name.

To accomplish these steps in Terraform, you'd start by creating the key:

```
resource "google_tags_tag_key" "c_key" {
    parent = "organizations/123456789"
    short_name = "contains_contracts"
    description = "To help grant contract reader access."
}
```

That numeric value would be your organizational ID. With the key created, the values would come next. Notice below the Terraform reference to the key's name. Remember, that's not the short name we are defining, it's the unique identifier name that Google will autogenerate on key creation:

```
resource "google_tags_tag_value" "c_value" {
    parent = "tagKeys/${google_tags_tag_key.c_key.name}"
    short_name = "true"
    description = "Project contains contracts."
}
```

With the key and value created, now you could bind the desired key/value to the requisite project, folder, or organization. By default, tags are inherited from parent to child in the resource hierarchy. The only way for a child resource to change this behavior would be to bind the same key with a different (or no) value.

When binding resources, the full resource name must be used. For a list of resource names, see https://cloud.google.com/asset-inventory/docs/resource-name-format. Here, I'm binding to a project:

```
resource "google_tags_tag_binding" "c_binding" {
    # The "join" below is a multi-line string hack:
    parent = join("", [
        "//cloudresourcemanager.googleapis.com/projects/",
        "some-project-id" #This can also accept a proj num
        ])
    tag_value =
        "tagValues/${google_tags_tag_value.c_value.name}"
}
```

Since the tag value name alone is unique, I don't need the key name to help identify the unique key/value pair.

Last, but not least, let's create a conditional IAM binding tied to our new pair. It's possible to create the condition on the namespaced name or the globally unique name. The namespaced name is easier to read, but if someone removes a key/value pair, and recreates a new one with the same name, you could end up with the same namespaced name only with a totally new meaning, and thus have legacy policies inadvertently keying off of an unrelated tag. Globally unique names are more random but never reused. When using automation and to minimize risk, it's probably a better practice to use unique ID names:

```
module "storage_bucket-iam-bindings" {
  source              = "…details omitted for space"
  mode                = "additive"
  conditional_bindings = [
    {
      role = "roles/storage.objectViewer"
      title = "contract_reviewer"
      expression = join("", [
      "resource.matchTagId(",
      "${google_tags_tag_value.c_key.name},",
      "${google_tags_tag_value.c_value.name})"
      ])
      members = ["group: grp-gcp-contracts-team@gcp.how"]
    }
  ]}
```

Excellent – so in addition to setting up conditions tied to things such as resource names, we can also build them for tags. Good job.

Now, let's take a small peek at a totally new IAM feature, deny policies.

Deny policies

At the time of writing, Google has just released in preview a new ability to create deny policies. Deny policies allow for the ability to create deny rules tied to selected principles, forbidding them from using a set of permissions no matter what IAM access policies stipulate.

Imagine, for example, you want to restrict the ability to create custom IAM roles to members of a particular team. With deny policies, you could define a rule that denies the role management permissions from everyone in the org who isn't part of the super special role team. The rule would look something like this:

```
{
    "deniedPrincipals": [
        "principalSet://goog/public:all"
    ],
    "exceptionPrincipals": [
        "principalSet://goog/group/cool-role-admins@gcp.how"
    ],
    "deniedPermissions": [
        "iam.googleapis.com/roles.create",
        "iam.googleapis.com/roles.delete",
        "iam.googleapis.com/roles.update",
    ]
}
```

You would wrap the above rule in a deny policy file and apply it with a new `--kind=denypolicies` switch on the `gcloud iam policies create` command. If you'd like to read more specifics, see `https://cloud.google.com/iam/docs/deny-access`.

With this rule in place, even if you were in the org administrator group and should have the ability to create custom roles, the deny policy would take precedence over your assigned IAM allow policies, and you would be unable to create custom IAM roles (at least until you added yourself to the allowed group).

Nice – that's one more way we can fine-tune our IAM assignments. Speaking of IAM assignments, let's take a moment to give some special attention to a subset of our user identities.

Limiting the use of privileged identities

Privileged identities, with the exception of those associated with our Terraform service account, should be tightly controlled and typically only accessed through some sort of firecall process. An example case might be that something in our Terraform CI/CD pipeline has broken down, the automation is no longer working properly, and you need to get in there with an account with an extremely high level of access to try and figure out what's happening.

The firecall process procedure should be well known and documented. To enforce limited use of privileged accounts, a requirement should be put in place that the approval of multiple individuals is required for the accounts to be used. One way of enforcing such a requirement would be to separate those who know the password from those who can access the required MFA token.

Principals should be considered privileged if they have any of the following roles:

- Super Administrator (group: `gcp-superadmin@`), with full control over Cloud Identity (`https://admin.google.com/`). Remember, this user is created directly in Cloud Identity / Google Workspace and doesn't come from the org identity provider or authenticate through standard SSO. It's great if there's a configuration issue with Cloud Identity / Google Workspace, but it also means it exists outside of typical controls related to your user management and identity provider systems.

- Organization Administrator (`gcp-orgadmin@`), with the ability to control the resource hierarchy, set IAM role bindings, and control org policies.

- Billing Account Creator (`gcp-billingcreator@`), with the ability to create new billing accounts for the organization, which are then administered by the following.

- Billing Account Administrator (`gcp-billingadmin@`), with the ability to manage anything billing-related from viewing usage and pricing to associating a given billing account with a project.

Now, before we put this chapter to bed, let's take a brief look at what to do when things go wrong with IAM permissions.

Troubleshooting access

If you ever run into a situation where a user or group seems to have more or fewer permissions than they should, troubleshooting can be trickier than you might think, at least without the right tools. As a result, all too often, users get assigned more permissions than they need, which can adversely affect a principle of least privileges approach to security.

Let's start with Policy Troubleshooter.

Policy Troubleshooter was developed by Google to help troubleshoot access issues. The process starts when a user attempts to visit a resource they don't have access to. The GCP console pops up a message resembling the following figure.

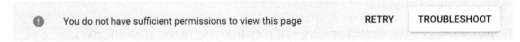

There was an error while loading /iam-admin/iam?project=sturdy-dogfish-330317

You are missing the following required permissions:

Project
resourcemanager.projects.getIamPolicy

Check that the folder, organization, and project IDs are valid and you have permissions to access them. Learn more

Send feedback

Figure 5.5 – Ready to troubleshoot

The message is telling my test user that she doesn't have access to view IAM settings, more specifically, that she's missing the getIamPolicy permission. If she hits the **TROUBLESHOOT** button, Google Cloud will forward her to the Policy Troubleshooter.

The problem for my user in this case is that she can see the Troubleshooter page, but she doesn't actually have the power to run the tool herself. But what she can do is copy the URL and pass it to her admin… Me! This is an important step if you are ever troubleshooting access for a user. You can talk the user through explaining what happened, and attempt to get the details that way, but it's so much easier if they just provide you with the URL.

When I drop the URL into my browser, it takes me directly to the Policy Troubleshooter all pre-filled with the user's information and exactly what permission they were missing, as shown in the following figure.

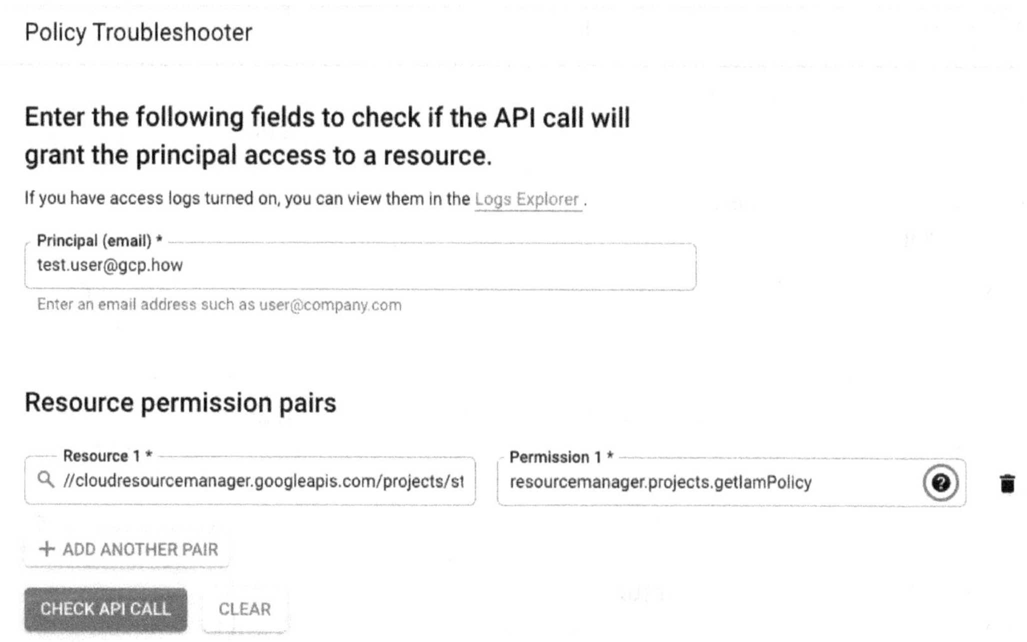

Policy Troubleshooter

Enter the following fields to check if the API call will grant the principal access to a resource.

If you have access logs turned on, you can view them in the Logs Explorer .

Principal (email) *

test.user@gcp.how

Enter an email address such as user@company.com

Resource permission pairs

Resource 1 *

//cloudresourcemanager.googleapis.com/projects/st

Permission 1 *

resourcemanager.projects.getIamPolicy

+ ADD ANOTHER PAIR

CHECK API CALL   CLEAR

Figure 5.6 – Policy Troubleshooter

You will see that the tool needs three pieces of information:

- The principal's email, which could be a user or group address.

- The name of the resource, again using the full technical names found here: `https://cloud.google.com/asset-inventory/docs/resource-name-format`. In this case, it's a project since my user was trying to view IAM permissions.

- The precise permission being tested.

When I execute the test, the page returned provides a hierarchal analysis of the permission and the principal. It displays what roles at the org, folder, and project levels possess the requisite permission, and which users and/or groups are bound into those levels of access. This way, if you think the user should have the permission because you thought they were in a particular group, it will let you know if a) the group has access, and b) whether or not the user is in the group. The tool will also analyze conditions. So, if the user should have access, but only 9-5 NYC time, the Policy Troubleshooter tool will allow you to easily determine so.

Another tool in this IAM troubleshooting section of the Cloud Console that can work in a related way is Policy Analyzer. The home page of Policy Analyzer prompts you with several pre-created queries for things such as *"Who can impersonate a service account?"* and *"What access does my employee have?"* If I select the what access query example, I can pick the scope from project to org. I'm going to go with org and drop in the principal for my test user. I can then generate a report of everything my test user has access to across the organization.

Let's say I'd like to get a list of everyone who can run queries using BigQuery in my billing logs project.

The Policy Analyzer provides a lot of good information on permissions by resource, principal, role, or specific permission, but remember with the resource and permission versions, you will have to do a little research so you can get the names precisely right (for an easier way, see the Cloud Asset Inventory tool). For BigQuery in the billing logs project example, I would need to determine the full resource path. If I research the document I've shared with you several times for the specific resource name, I see it has an example for access to a particular dataset, but it turns out the resource list isn't complete. If instead, you type `bigquery` into the resource field in the Policy Analyzer, it will present you with a list of options. With a little experimentation you will find something like the following:

```
//serviceusage.googleapis.com/projects/<project numeric ID>/
services/bigquery.googleapis.com
```

Since it requires the project's numeric ID, I do a quick trip to the billing logs export project itself and on its home page, I see the numeric ID listed. I drop that into the resource name and when I run my query, I can see exactly which users and groups have the ability to access BigQuery within the project, and why exactly they have the access. In the details, you can even check the Policy History and see changes to permissions within the last 35 days.

Policy Analyzer is actually part of one final tool I'd like to mention from the IAM and Admin section of Google Cloud, Cloud Asset Inventory. Google's Cloud Asset Inventory tool can help you find assets by metadata, export asset metadata at moments of time, examine changes over a time range, receive real-time notifications of asset changes, and analyze IAM policies to find out which principal has access to what resource.

In this discussion, I'm going to focus on how Cloud Asset Inventory helps with IAM Policy analysis, and how it links back to the aforementioned Policy Analyzer.

The home page of Asset Inventory displays a curated list of assets, projects, and locations. Clicking on the **RESOURCE** tab will give you the opportunity to search by resource.

As an example, say I want to know who exactly has access to my Terraform state bucket that I set up when I bootstrapped the CI/CD and seed projects. In the Asset Inventory Resource search box, I could search `storage`. Not only do I get a list of buckets and other resources using the name `storage`, but I also see the asset types on the left side. Checking **storage.Bucket** will give me just the list of buckets. You can see the results in the following figure:

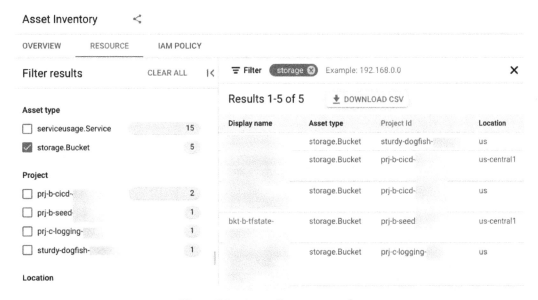

Figure 5.7 – Assent Inventory results

Clicking on the state bucket will pop up details. Clicking on any of the details, such as location or project ID will add those fields to the search filters. Clicking over to the **IAM POLICIES** tab will display all the principals with access to the selected bucket, and which roles they get that access from.

Principal ↑	Roles
group:gcp-organization-admins@gcp.how	roles/storage.admin
projectEditor:prj-b-seed-	roles/storage.legacyBucketOwner,roles/storage.legacyObjectOwner
projectOwner:prj-b-seed-	roles/storage.legacyBucketOwner,roles/storage.legacyObjectOwner
projectViewer:prj-b-seed	roles/storage.legacyBucketReader,roles/storage.legacyObjectReader
serviceAccount: '@cloudbuild.gserviceaccount.com	roles/storage.admin
serviceAccount:org-terraform@prj-b-seed- iam.gserviceaccount.com	roles/storage.admin

Figure 5.8 – State bucket IAM POLICIES

A good way to get more details is to close the search results and navigate to the Asset Inventory **IAM Policy** tab. There, if I check the box to filter by the same **storage.Bucket** type, I will see right in the results the principals and roles for my various buckets. If I again click into my state bucket details, I will get a list of roles, with principals for each role, as you can see in the following figure:

Binding details

Role	Principal	
▼ Storage Admin		
	⋮⋮ gcp-organization-admins@gcp.how	ANALYZE FULL ACCESS
	⊡ @cloudbuild.gserviceaccount.com	ANALYZE FULL ACCESS
	⊡ org-terraform@prj-b-seed- iam.gserviceaccount.com	ANALYZE FULL ACCESS
▼ Storage Legacy Bucket Owner		
	⋮⋮ Editors of project: prj-b-seed-	ANALYZE FULL ACCESS
	⋮⋮ Owners of project: prj-b-seed-	ANALYZE FULL ACCESS
▼ Storage Legacy Bucket Reader		
	⋮⋮ Viewers of project: prj-b-seed	ANALYZE FULL ACCESS
▼ Storage Legacy Object Owner		
	⋮⋮ Editors of project: prj-b-seed-	ANALYZE FULL ACCESS
	⋮⋮ Owners of project: prj-b-seed-	ANALYZE FULL ACCESS
▼ Storage Legacy Object Reader		
	⋮⋮ Viewers of project: prj-b-seed	ANALYZE FULL ACCESS

Figure 5.9 – State bucket role list

Clicking **ANALYZE FULL ACCESS** for any of the choices will take you back to the Policy Analyzer where you can modify or simply run the query that was generated for you by the Asset Inventory tool.

So, If I wanted to know who could run queries over the dataset where my billing logs are being exported, I could do the following:

1. Launch the Asset Inventory tool and switch to the **IAM POLICY** tab.

2. Filter to the billing logs project.

3. In the results, there will be a link to the datasets in the project. Click on the one where billing data is being exported.

4. Examine the principals with access, like in the following figure:

Binding details

Role	Principal	
▼ BigQuery Data Editor		
	⠿ Editors of project: gcp-how-billing	ANALYZE FULL ACCESS
	⠿ service-' '@gcp-sa-bigquerydatatransfer.iam.gserviceaccount.com	ANALYZE FULL ACCESS
▼ BigQuery Data Owner		
	⠿ Owners of project: gcp-how-billing	ANALYZE FULL ACCESS
	⠿ billing-export-bigquery@system.gserviceaccount.com	ANALYZE FULL ACCESS
	⠿ patrick@gcp.how	ANALYZE FULL ACCESS
▼ BigQuery Data Viewer		
	⠿ Viewers of project: gcp-how-billing	ANALYZE FULL ACCESS

Figure 5.10 – Access to billing exports

Great stuff, and with that, it's time to wrap up our access discussion and move on to see what our Google Cloud foundation has in store for us next.

Summary

In this chapter, we continued to build a secure and extensible foundation in Google Cloud by completing step 6 and added IAM trust boundaries to our resource hierarchy. We learned how Google Cloud IAM works, and how to select appropriately from the three IAM role types. We then learned how to plan access control in our resource hierarchy and how to use Terraform infrastructure as code automation to implement our plan. Lastly, we learned how we can troubleshoot access problems as they arise.

Fantastic job y'all, really!

If you want to keep on keeping on through Google's ten-step checklist with me, your personal tutor, by your side, flip the page to the next chapter where it will be time to get to building our Google Cloud VPC network.

6
Laying the Network

Now and again, my boss sends me to lead large, *"what's cool about Google Cloud,"* events. The audience is always excited; excited about the technology and excited to be around others in the industry doing the same thing they are. And I'm an excitable guy, and all that excitement in a big room builds energy all its own. Typically, there are too many people to allow questions during the event itself, but at the end, I like to open the floor and do a Q/A session.

I was teamed up with some people from Google one time doing a gig in Dallas, TX. We were way up high in this hotel ballroom with a spectacular view out of a floor-to-ceiling side window. When the time for Q/A arrived, a young lady stood up and said, *"I'm about to graduate with a degree in computer science and my specialty is networking. After listening to your talk, I feel like I might have picked a bad specialization. Will I be able to get a job in the era of the cloud?"*

For a moment I stood there, likely with my mouth hanging open, silently wondering how I'd ever given her that impression. Get a job in cloud? With a college specialization in networking?

"Heck yeah, you can get a job in cloud!"

I mean, have you seen Google's official list of Google Cloud Professional Certifications? If not, have a look at `https://cloud.google.com/certification`. At the time of writing, Google has 10 professional-level certifications, and two of those certifications have a lot of networking-related questions: Cloud Network Engineer (!!!) and Cloud Security Engineer.

Networking is at least as important in Google Cloud as it would be in an on-premises data center. Granted, you aren't going to be setting up switches and pulling network cable, since that's part of the low-level infrastructure Google has created for their cloud users, but you are going to be making a lot of networking-related decisions, and those decisions are going to have a large impact on the security and extensibility of your cloud presence.

Great – with the importance of Google Cloud networking established, let's get to work building our foundational Google Cloud network.

In this chapter, we are going to discuss some key elements in Google Cloud networking, make some key decisions and plan our network, and see how Terraform could help implement its design. We will be covering the following topics:

- Networking in Google Cloud
- Step 7 – building and configuring our foundational VPC network

Networking in Google Cloud

At its most basic level, networking is used to connect multiple systems. I'm typing this on my main work Mac at home. There's a wire running from my machine to a new switch I just installed downstairs as part of networking upgrades at my home. The switch brings my various wired devices and wireless access points together so that everyone in the house can get online. Behind that, a wire runs to a router that joins my two internet providers. It also provides a lot of network services, including DHCP, firewalls, **Network Address Translation** (**NAT**) to give me a public IP when accessing the internet, denial of service attack protection, and so forth. My router then has a couple of wires connecting it to my two-internet provider modems/routers. Each of them offers more protection and networking services. Lastly, there's a controller device connected to the switch to make everything work as a unit. That's where I can pull up a web page and create all my configurations and settings.

Cool, right? And if I ever decide to set up physical networking again, I'm going to have my head examined. Do you have any idea how much research I had to do to make all that crap work? 2 weeks ago, I didn't know what gigabit + multi-WAN routers with load balancing and failover were, let alone how to configure them. There's an adage that says: beware the tool-wielding software developer, for there is a no more dangerous animal. I resemble that remark.

However, what was interesting about the entire experience was how well it highlighted the differences and similarities between physical and software-defined networks. In Google Cloud, you (or your grandson, in my case) don't have to pull network cables through the attic, nor set up physical routers and switches, because Google has built and manages all the physical network components for you. What you do in GCP is define a **software-defined network** (**SDN**), which Google calls the **Virtual Private Cloud** (**VPC**).

> **This Chapter Could Be a Book by Itself**
>
> The point I was trying to make with my opening story in this chapter, with the new college networking expert grad, is that networking is a hugely complex topic in Google Cloud. If you aren't strong in the networking arts, then make sure someone you work with is willing to put in the time to learn more about networking in Google Cloud.

Understanding Virtual Private Cloud networks

There's a nice quote from *Wired* magazine that states *"This is what makes Google Google: its physical network, its thousands of fiber miles, and those many thousands of servers that, in aggregate, add up to the mother of all clouds."*

If you head over to `https://cloud.google.com/about/locations#network`, you can see the current view of the following diagram:

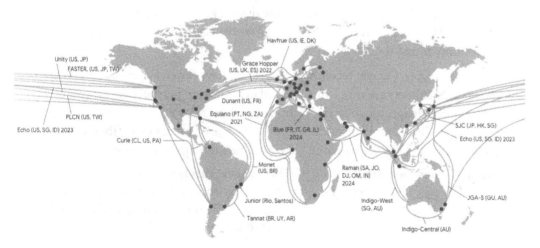

Figure 6.1 – Google's physical network

I say current because, like its data centers, Google is constantly expanding its network. Google believes that only having a fast and reliable network will allow the continued use and expansion of cloud computing, and at this point, I believe they have the largest privately-owned network in the world. The black dots? Those are the Google Cloud **points of presence** (**POPs**), where the Google network meets up with the rest of the world.

But that's physical networking. We need to talk about the SDN that we use as Google Cloud clients. Let's cover a little terminology:

- **Virtual Private Cloud** (**VPC**): A global-level box that represents the logical unit of isolation that is your network. When you think *"My machine is on this network, and out there is the rest of the world and other networks,"* that's the analog to your virtual machine being attached to a particular VPC. Barring some special type of connection (peering or VPN), only public IPs are accessible outside a given network.

- **Subnet**: A region-specific, logical group of private IP addresses. While a Google Cloud VPC has no native IP address range, the subnets do. The router or modem on your home or office network likely provides you with the private IP you are currently using. Google Cloud subnets do the same thing for attached resources.

  Google reserves four IPs in each subnet for their internal use. As an example, imagine that we have created a subnet using the 10.0.0.0/24 CIDR range. Here, Google would reserve the first IP in the range, which represents the network – that is, 10.0.0.0. The second IP works as the default gateway – that is, 10.0.0.1. The second to last IP is reserved but not currently used – that is, 10.0.0.254. The last IP is broadcast: 10.0.0.255.

  As a best practice, Google recommends that you have fewer, larger subnets. While app-specific subnets are nice from a logical organization perspective, they can be tough to manage, tough to size right, and you are allowed a maximum of 300 subnets per VPC, which means that if you have lots of small ranges, then you are going to hit that limit and run out. Larger subnets tend to work better.

- **Classless Inter-Domain Routing** (**CIDR**) **ranges**: This is a way to specify a block of IP addresses. CIDR ranges take the form of *network address/bitmask*, such as 192.168.0.0/24. I'm sure we could have a whole discussion on CIDR, and IPv4 versus IPv6 ranges, but I'll leave that research to you.

- **Firewall rules**: These allow you to control *which* packets are allowed in and out of *what* networked resources. Each Google Cloud VPC implements micro-segmentation firewall rules. Many physical networks implement segmentation firewalls, typically with hardware or virtual appliances, to separate the internal network from systems in the outside world. Since a Google Cloud VPC is a software-defined network, it can offer software-defined firewall rules that essentially evaluate the packets at the edge of each networked resource (micro-segmentation) rather than simply at the network's access point. So, it's like all the firewalls are sitting just outside every VM (micro-segmentation), rather than only at the access point to the network itself (segmentation).

- **Routes**: These tell networked resources how to send packets from the source instance to the destination. The VPC will come with auto-generated routes for all the subnets on the network and allow traffic in and out of the network, but you can add manually created static routes and/or dynamic routes that are managed by Google's Cloud Router.

- **Cloud Router**: This allows you to dynamically exchange routes between multiple individual networks using the **Border Gateway Protocol** (**BGP**).

- **Cloud DNS**: This is a GCP-based, fully managed DNS service that offers both public and private DNS zones, all of which are managed through a central DNS hub.

In Google Cloud, each VPC network, like any GCP resource, will be owned and managed by a single project. Google offers three basic VPC configurations (two types of networks, one with two types of subnets):

- **Default (don't use)**: The default network is created anytime a new project is created. My "don't use" recommendation isn't because the default network is inherently bad; it's more that it makes some assumptions that may not be in line with your networking goals. For example, the default network comes out of the box with a single /20 subnet in every region of the world. Look at a CIDR table such as `http://www.rjsmith.com/CIDR-Table.html` and you'll see that /20 is about 4,000 IPs. Is that too many? Not enough? And besides, are you going to put networked resources in every GCP region? If you don't have resources in a region, don't put a subnet there.

  The default network also comes with four default firewall rules. The first three allow RDP, SSH, and ICMP (think ping and traceroute) access to any machine on the network, while the fourth allows all the machines to communicate with each other. Allow external SSH and RDP access to all my machines? I think not.

My favorite default network quote is from the Google Cloud documentation itself, where it says that the default network "*is not suitable for production environments.*"

Who am I to disagree with Google?

- **Custom network with automatic subnets (don't use)**: Essentially, this is the same thing as the default network, but you get to create the network first and give it a unique name, and then Google creates all the subnets automatically. The same issues arise for me here as those in the default network. The fact that the name is something other than the default doesn't mean the network architecture is any better.

- **Custom network with custom subnets (ding ding ding, we have a winner!)**: Using a custom network with custom subnets is Google's recommended best practice (`https://cloud.google.com/architecture/best-practices-vpc-design`). Essentially, Google's recommendation is to rip out the default network (or set a policy so that it never gets created in the first place) and then go back, create a custom network, and manually create the subnets and firewalls you will need. That's an oversimplification, but at least it's in the right ballpark.

A simple custom VPC with a subnet in two zones may look something like what's shown in the following diagram. Note that since there's no CIDR range associated with the network as a whole, so long as the subnet ranges don't overlap, any valid IP range (`https://cloud.google.com/vpc/docs/subnets#valid-ranges`) is allowed for each subnet:

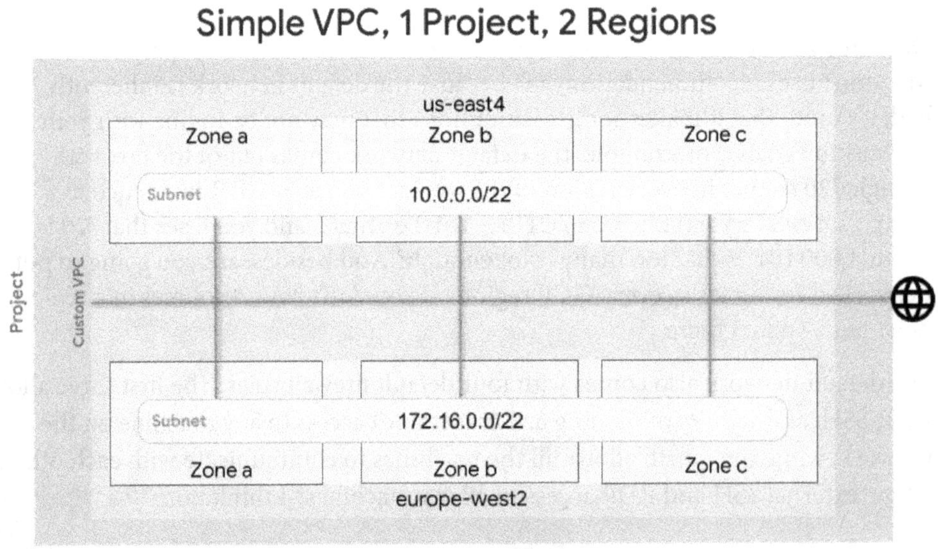

Figure 6.2 – Simple VPC, 1 Project, 2 Regions

Perfect! Now that we have a basic understanding of the major components comprising Google Cloud VPC networks, let's look a bit more closely at how networked resources communicate.

Communicating between networked resources

Several GCP compute technologies can operate on the VPC, including Compute Engine, Kubernetes Engine, and serverless VPC connectors where support can be extended to serverless products, including Cloud Run, Cloud Functions, and App Engine.

So, imagine starting with a project where you've removed the default network and added the custom VPC network with two subnets, as described previously. You add a VM to each of your subnets, in each of your regions. You ask the `us-east4-a` machine for its IP and it returns `10.0.0.2` (remember, the first two IPs in the subnet are used by Google). You ask the `europe-west2-a` machine for its IP and it returns `172.16.0.2`. Since the VPC itself has no master CIDR range, the fact that you used two different base ranges for your two different subnets isn't a problem. Assuming the firewalls allow it, and since the two machines are on the same network, they can freely communicate using their internal IPs.

> **Surprise – Networking Can Cost Money**
>
> Make sure you spend a little quality time reading Google's network pricing page: `https://cloud.google.com/vpc/network-pricing#vpc-pricing`. At the time of writing, two VMs *in the same zone* and *on the same network* communicate for free, but machines in different zones, even if those zones are in the same region, are charged a network egress fee. If said machines are in different regions, then they are charged a fee, depending on the source and destination locations. Data downloaded out of the cloud to external systems, no matter which resource it comes from, also costs money. Don't design a network without thinking about the cost ramifications for said design.

If two Compute Engine machines aren't on the same network, and there's no special connection between the two different networks (peering, VPN, and so on), then the only way they can communicate is using publicly routable external IPs. The packets would leave machine one's VPC, head out into the Google Cloud edge network, and then run through Google's edge network security, back down into the second VPC, and then to its VM. That trip up and back, into and out of Google's edge network, adds overhead, latency, and cost.

> **Note – By Default, Compute Engine VMs Have Both an Internal and External IP**
>
> The internal IP is the machine's actual IP and it's allocated from the VPC subnet that the machine is attached to. Internal IPs are private network IPs and are not routable outside of the VPC. The external IP will be used for exposing the resource to the outside world and is routable outside the VPC. External IPs essentially act as alias IPs – that is, traffic that's pointed at the external IP is simply routed by Google to the internal IP. To minimize the surface area you expose to attacks, external IPs should be removed wherever possible. Later in this book, we will learn that we can mandate this practice for all or part of our organization using an org policy.

Now that we know how communication works within a VPC and between VPCs, let's learn about special VPC relationships.

Connecting VPC networks

In the previous section, we discussed how the only way your VMs can talk to my VMs would be to use external IPs. Well, that's not completely true. Another approach might be to set up a special trust relationship between our mutual VPCs.

Other than networks communicating with external IPs, there are three ways VPCs can be connected so that routing traffic using private IPs is possible: peering, using a VPN, or using a multi-NIC VM instance attached to both networks. Let's start with the simplest: peering.

VPC peering

VPC network peering allows two VPC networks to be connected, across projects, and even across organizations. The two networks, instead of acting like two trust boundaries, operate as a single unit with a trust boundary that encompasses everything in both networks. Resources can communicate via a private IP address and the packets never move into the edge region of Google's network.

Peering VPCs has several advantages over the other ways of connecting networks. Since the packets move across Google's internal network and not through the edge network, the latency is lower, the resources don't have to be exposed via an external IP address helping with security, and the networking cost will be the same as if the networked resources were on the same network, possibly saving money.

To summarize, the two VPCs operate just like a single larger VPC.

Peered networks are still administered individually, by the respective network owner projects, so custom routes, firewalls, VPNs, and the like will need to be configured on either side. Initial peer setup will require action on both sides and the connection can also be deleted from either side.

> **Note – There's an Implicit Deny Ingress Rule You Will Need to Address**
>
> VPC networks all have two implied firewall rules, neither of which appear on the Console firewall page. The implicit *allow egress* rule is the lowest priority firewall and allows packet traffic not addressed to an explicitly routed IP to leave the network. The lowest priority implicit *deny ingress* rule will block all traffic from outside a VPC network from entering it. Since the two VPCs are still two different VPCs, each with its own firewalls, the implied deny ingress rule will block traffic from entering the VPCs, even via a peered connection. You will need to explicitly create firewalls on both sides to allow access. Having said that, the firewall rules can also grant limited access, so perhaps VPC1 can only access particular subnets in VPC2.

Only directly peered networks can communicate, and transitive peering is not supported. If networks A and B have been peered, and networks B and C have also been peered, network A cannot communicate with network C unless it is also directly peered.

There are two major downsides to network peering. First, there is no way to control which subnet routes get shared. If network A has a subnet that's using `10.0.0.0/22`, then network B can't use that subnet, since the two networks will be operating as a single network, within a single IP space. Though you can't control how subnet routes are shared, you can control how custom routes are exchanged, so keep that in mind.

The second, and arguably more major issue with peered networks, are their limits. Make sure that you review the peered network limits documentation here: `https://cloud.google.com/vpc/docs/quota#vpc-peering`. Some key limits are as follows:

- A network may have up to 25 different peered connections
- There's a maximum of 300 static and 300 dynamic routes
- No more than 15,000 VMs are allowed in a peering group
- A maximum of 400 subnets across the group

If the downsides don't get in your way, then peering is typically the best, and easiest, way of setting up special VPC relationships. You get administrative independence, higher security, and lower latency, all for the same cost as networking on a single VPC. Not bad.

Now, if VPC peering quotas and limits are tripping you up, then the next best way of connecting networks is to use a VPN.

HA VPNs

The first time I ever came across Google Cloud VPNs was in an article discussing various ways to integrate an on-premises network with a Google Cloud-based VPC. You have a VPN gateway appliance in your on-premises data center, and you connect that gateway to a Cloud VPN in Google Cloud. The configuration essentially extends the on-premises network to the Google Cloud VPC. That's cool, and we'll talk more about how to use VPNs like this later in this chapter, but that's not the use case now. For now, we want to connect two VPC networks, both in Google Cloud.

Before we go any further, let's discuss some cloud VPN negatives. A VPN has a theoretical maximum throughput of 3 Gbps (1.5 up and 1.5 down), but the actual bandwidth with overhead and latency is typically closer to 1.5 Gbps. Where peering is global, VPNs are regional resources. The VPN connection would have to be built in a single GCP region. VPNs are going to communicate via external IPs, which as we discussed earlier, is going to add latency and cost, as the link is going to run through Google's edge network.

That's a lot of negatives, so what's the upside? Why would you want to use VPNs to connect Google Cloud VPCs?

The short answer? VPC networks that are connected using VPNs don't share limits and quotas. So, if you need to connect two VPCs, even those that belong to different organizations, but the peering limits mentioned earlier are going to be a problem, and if the throughput offered by a or multiple VPNs will fit with your use case, then consider setting up a HA VPN connection. One other upside of VPNing your VPCs is that the connections can be transitive if needed.

HA VPN Versus Classic VPN

In the bad old days, Google Cloud VPNs offered the ability to do static routing (as opposed to auto-sharing routes with Cloud Router and BGP), and they supported connecting through a single VPN tunnel. In early 2022, Google deprecated parts of Classic VPNs to shift all Google Cloud VPNs to HA VPNs. In HA VPNs, there are always pairs of tunnels instead of singles, and all subnet routing is handled via BGP. If you are currently working with a Classic VPN, make sure to read about the changes occurring and pencil in "move to HA VPN" on your to-do list.

VPC peering and HA VPN connections represent the majority of use cases for connecting VPCs. The third choice I wanted to mention is viable but specialized.

Multi-NIC VM instance connections

In Google Cloud, a VM can have multiple (up to eight, if you have at least 8 CPUs) network interfaces. Each NIC may then be connected to a subnet in a different network. The VM can then act as a bridge across the multiple networks. The connection would be tied to the region and zone where the VM lives, and the cost and performance would depend on the size and capabilities of the VM.

> **Compute Engine Network Throughput**
>
> In Google Cloud Compute Engine VMs, network throughput is directly proportional to the number of CPUs on the VM, up to a maximum of 32 Gbps at 16 CPUs. In early 2022, Google introduced a new feature called Tier 1 bandwidth, and now on N2, N2D, C2, and C2D machine types with 32 CPUs or more, Tier 1 high bandwidth can deliver 50-100 Gbps, again, depending on the number of CPUs, for an extra fee of course.

There are two main use cases for bridging networks through a VM. The first is to allow the VM to run some sort of network-related software, creating a software-driven network appliance that can then apply additional security, monitoring, and control to the traffic moving through the VM.

The second common use case is to use a VM to help with perimeter isolation. One of the NICs can face the outside world, while the other faces internal resources. GCP firewalls can be used to limit traffic in and out of each interface. This approach can be used for something such as an externally facing web service, where incoming traffic runs through the external NIC, while anything the app needs to access inside Google Cloud (app servers, databases, and so on) runs through the internal-facing controller.

Excellent! Now that we know a bit about GCP networking in general, as well as how to connect networks, let's dig into a way to expand a single network so that it's larger than just the one host project.

Leveraging Shared VPCs

As I've mentioned previously, I'm a huge fan of the KISS principle and the idea of building a Google Cloud environment using single networks owned by single projects. Sounds so... simple, right? At least, until you think about things.

Imagine this: you are about to start work on a new *cool app*. You start sketching a design where you will create a `cool-project` in which you can create a `cool-vpc`, with a corresponding `cool-subnet` in the region where the service will live. Then, you can build your resources in the project and put them on the VPC. Easy, right?

Wrong. It only sounds easy because you're looking at a tree and haven't considered the rest of the forest.

Think about it. What's our organization's structure going to look like with the new cool app in it? Something like this:

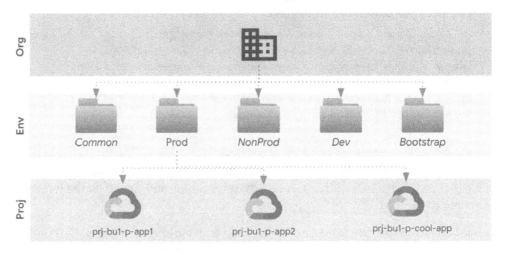

Figure 6.3 – Cool App

If you're following my recommended design, the first problem you're going to encounter is the different environments. You're going to need to have a dev, non-prod, and prod version of the *cool-app* network, and you're going to need to keep an eye on your automation to make sure they are all configured appropriately.

OK, it's not terrible, but next, I want you to think about the forest. Take the stated design and extrapolate it out to all the different apps you could conceivably create. Oh, and I'm going to assume that some of your apps are going to want to talk to each other, yeah? Are you going to do that all via external IPs? As we saw earlier, that could work, but it would require firewall configurations, not to mention the extra latency and networking costs. Want to go the VPC peering route? Well, not only will all those peered connections end up looking like a spider web, but you also need to remember the 25 network limit on peering, because I bet we could run up against that real fast.

It sounds to me like we're headed toward a tough-to-manage mess. I'm not saying it can't be done. I can tell you with all honesty that I've cringed upon seeing it in production more than once, and seriously, we can put our college student from earlier in this chapter to work, doing nothing but trying to keep all the networks working properly. She's not going to keep working for you for long, however, because she's going to see right off that there's a much better way, and when you ignore her advice for long enough, she's going to find a better job.

Without further ado, allow me to introduce you to the Shared VPC. **Shared VPC networks** allow you to create and manage the VPC and its subnets, firewalls, and routes all within a central **host project**. The host project centralizes management, so you can do your network configurations once. Multiple **service projects** may then be attached, and they all share the single, centrally managed, Shared VPC network. IAM permissions will allow for fine-grained control over exactly who can do what in both the host and service projects, giving you a much better way to apply a least permissions approach, all without compromising the mental equilibrium of your network administrator.

So, a Shared VPC is simply a VPC that's shared across multiple projects. Nice.

A Shared VPC can share all of its subnets with all the service projects, or you can be more selective and share specific subnets with specific service projects. The nice thing about the latter is that you get more control over which subnets get used where, and who can access each. If you expect the *cool-app-project* to use a particular subnet or two, give them access to only those subnets.

To create a Shared VPC, a user or group needs to be granted the Shared VPC Admin and Project IAM Admin roles. With that, the user can go to the **Shared VPC** page in Google Cloud and choose **Set up Shared VPC**. Of course, you're not going to do it that way, because you are going to use Terraform (right?), but you could. If you'd like more details, go to `https://cloud.google.com/vpc/docs/provisioning-shared-vpc#setting_up`.

When you create a Shared VPC host project, the project's network isn't automatically shared. To do that, you will have to attach service projects. Each service project may only be attached to a single Shared VPC, but Shared VPCs can be peered.

The Shared VPC Admin can assign a Service Project Admin in each service project and grant that admin access to all, or a subset of, the subnets in the host project. With that done, the Service Project Admins can create resources in the service project by using subnets in the host project.

Excellent! Now that we know how to build networks, let's discuss a few related networking topics, starting with hybrid cloud connectivity.

Hybrid cloud options

When Google says the words **hybrid cloud**, what they mean is an IT infrastructure where part of it is running in your on-premises data centers, and part of it is in Google Cloud. Why not put all of it in the cloud? Well, the key reasons tend to be as follows:

- **Speed of light**: No matter how close you are to the GCP region you are using, it will never be as low latency as a server in the same building as you. If you are processing a bunch of data on-premises, then having the data right there with the processing will make the most sense.

  Go read about CERN sometime. When they turn on all the data collectors at their particle collider, they are collecting 1 PB of data per second. You could never send that all to GCP – not that fast. What they do is collect it locally, do a bunch of data transformations and aggregations, and then they send the resultant (much, much smaller) chunk of data up to Google Cloud to be studied long term.

- **Cost**: If you've got a 10-year-old piece of software already slated to be replaced, and it's running on a 5-year-old server, then it's practically running for free. Moving it to the cloud may not be possible, based on the type of application, and even if it is, it's likely to cost you more money running it there than leaving it where it is on-premises – until you can finally get rid of it.

  Another cloud cost to consider is networking. Remember, loading data into GCP is free, but getting it out costs network egress. Even if your latency to GCP is minuscule, storing a ton of data in GCP, then processing it on-premises or vice versa could cost you a lot at the network level.

  In my CERN example, the raw data storage cost would be another factor regarding why they don't upload it to GCP right off the bat.

- **Regulatory compliance**: I got called in to offer advice one time on a gig where a company in New Zealand was planning a big app they wanted to build in Google Cloud. I'm in this Zoom meeting (because they didn't want to fly me down, damn it), tossing out ideas and suggestions, and I asked, "*Are there any regulatory compliance issues with us moving this data out of New Zealand?*"

  One of the clients says, "*Oh yeah, for sure. This data can't leave New Zealand at all.*"

  The call went quiet at that point, because Google Cloud currently doesn't have a region in New Zealand, and the plan up to that moment was to export all the data to Australia. After the meeting? Hybrid cloud, baby.

Google strongly believes that the future of IT is the hybrid cloud. If you are an organization planning to mix IT resources running in Google Cloud, and other resources running in your on-premises data centers, then how exactly you connect the environments becomes a big question.

In general, there are three ways you can connect the environments. The first connection option is the most obvious – just use the internet. If you have a decent network connection with bandwidth to fit your needs, then a standard internet connection can work fine. The only real drawback is the need for the connection to run from external IP to external IP. Exposing things using external IPs means you have a larger surface area to attack each side of the connection. We do that sort of stuff every day, but if you can work out another solution, do.

To help with security, a better design would not involve those external IPs. In other words, use one that extends the trust perimeter around the on-premises network so that it encompasses the GCP VPC network.

There are two (2.5?) ways to connect your on-premises network to Google's network and do so in such a way that you can leverage internal IPs: VPN or Interconnect.

We discussed HA VPNs earlier in this chapter as a way to connect different Google Cloud networks, but it can work just fine to connect an on-premises network to a Google Cloud network. You can set up a pair of VPN appliances on-premises (or a single appliance that supports multiple tunnels) and a pair of HA VPN tunnels in a GCP region, and then throw in a Cloud Router to handle the BGP route exchange. Once you get it all configured, you can get that 1.5-3 Gbps throughput all with a 99.99 SLA. You are still connecting over the internet, but the VPN allows for secure traffic to be encrypted and passed over the internet's insecure medium. Thus, the VPN allows for internal IPs to be used on either end of the connection and, once again, BGP ensures that the subnet routes for both environments are shared properly.

Another option is dedicated Interconnect. Dedicated Interconnect is where you run fiber from your data center to one of Google's points of presence. If you look back at *Figure 6.1* in this chapter, all those black dots are points of presence. Once Interconnect has been configured, it can use private IPs because you've connected your network to Google's. A variation of dedicated Interconnect is partner Interconnect. Instead of you running fiber to Google, you pay a third party to hook you up. The third party (`https://cloud.google.com/network-connectivity/docs/interconnect/concepts/service-providers`) is connected to Google, and you are connected to the third party. Where dedicated Interconnect grants you an SLA directly from Google, make sure you understand how (if) the SLA works between you, the third-party provider, and Google.

Excellent! Now that you know the basics of Google Cloud networking and how to connect your on-premises network to Google's, let's talk a bit about a really important topic: security.

Google Cloud network security

In general, security in Google Cloud is accomplished by building an in-depth defense that's created by layering multiple security measures. It's like in our office building example from an earlier chapter. If we needed our new offices to be highly secure, then we wouldn't have a single security layer at the front door and stop with that. We'd need to combine that initial front door security with added layers, perhaps at the floor level, department level, and sometimes even at the door to individual offices level. And that's all before we add extra security to individual systems. On top of that, we may add active scanning to track where individuals go and make sure that everyone has a badge appropriate to where they are. This layered approach is exactly what we want to do in our Google Cloud foundation.

I was once asked, "*If you have good application security, why bother adding network security? Isn't it just more to manage?*" It is more to manage, but when done right, not only does it increase security through layering, but it also reduces the blast radius if – and when – breaches occur.

Before I get carried away, let's keep our security discussion close to our current major topic: networking. Network security in Google Cloud has several major elements, above and beyond the architecture of the network design itself. Let's get a taste of those major measures, and then put together a plan that incorporates everything we've learned.

First, there are firewalls.

Limiting access with firewalls

As I mentioned previously, since a VPC is a software-defined network, one of the many services it offers is micro-segmentation firewalls. Firewall rules allow you to watch for specific traffic types and then, based on that traffic's protocol, port, source, and destination, the rule can decide whether to let the traffic pass through the firewall or not.

Each firewall rule has several major configurations:

- **Logs**: Enable or disable (default) the logging of firewall rule actions. We will talk more about firewall logs in *Chapter 7, Foundational Monitoring and Logging*.
- **Network**: The network the firewall will be associated with.

- **Priority**: 0 - 65,535, with 0 being the highest, 1,000 being the default, and 65,535 being the lowest priorities. In the case of conflicting firewall configurations, the highest priority rule takes precedence.

- **Direction**: *Ingress* for traffic coming in through the firewall and *egress* for traffic exiting the resource. Though ingress firewall rules are the more common, several use cases will require you to limit outbound traffic (PCI DSS compliance comes to mind).

- **Target**: Resources that the firewall will be applied to. Options include all the instances on the network or specific instances that have been identified by a network tag or service account, with service accounts being preferable.

- **Source**: Filter for ingress or egress rules. For example, allow traffic but only if the source is in a specified CIDR range.

- **Protocols and ports**: The type of traffic and destination port of traffic being filtered. For example, blocking TCP traffic targeted at port 22 would block SSH access to VMs.

- **Enforcement**: Whether the rule should be enabled or disabled.

As we mentioned earlier in this chapter, all VPCs have two implicit firewalls that do not appear in the Console. **Implied allow egress** is the lowest priority (65,535) allow rule for 0.0.0.0/0, which lets any instance on the network route traffic to any IP outside the network. **Implied deny ingress** is the lowest priority (65,535) deny rule that's attached to the source, 0.0.0.0/0, which blocks any traffic from outside the network from accessing anything in the VPC.

In addition to project-level firewalls, Google also offers hierarchical firewall policies. Hierarchical firewalls can be created at the folder or organizational level and can then be associated with the org or any folder in the resource hierarchy. From there, they will be inherited top-down, from the associated org/folder down to the instance. They can also be restricted to specific VPCs and instance service accounts.

Hierarchical firewall policies may be chained so that a policy at, say, the org level could grant specific traffic, and then let subfolders decide whether to continue to grant said access or block it.

Another key element in our network security is VPC Service Controls.

Mitigating data exfiltration risks with VPC Service Controls

VPC Service Controls help create logical perimeters to protect resources and data above and beyond what can be done just with IAM. Where IAM provides a granular identity-based access model, Service Controls enable context-based perimeter security, adding another layer of protection to our in-depth protection.

VPC Service Controls augment your security in several key ways, as follows:

- They logically isolate your resources and networks into service perimeters, allowing you to control what can be done inside the perimeter, and differentiating that from what can be done outside.

- They may be extended to include on-premises networks, making the permitters extensible for hybrid cloud use cases.

- They can limit access to resources and data from the internet.

- They can protect data exchanges across permitters and between organizations.

- They provide context-aware access control for resources tied to client attributes and ingress rules.

VPC Service Controls work for many, but not all, GCP resources. For a list of supported products, along with some product-specific limitations, go to `https://cloud.google.com/vpc-service-controls/docs/supported-products`. For details on how VPC Service Controls are implemented, go to `https://cloud.google.com/vpc-service-controls/docs/overview`.

Good examples of how VPC Service Controls augment security are as follows:

- They prevent private data from being exposed accidentally due to misconfigured IAM policies. Let's say that someone miss-sets the permissions on a bucket so that non-employees can access the files. However, the perimeter has been configured so that if you're not in GCP or on the corporate network, then you can't move data outside the perimeter.

- They prevent data exfiltration beyond the perimeter by hacked code or compromised insiders. The code or insider may officially have access, but when they try to move the data beyond the Service Control perimeter, they are blocked.

- They prevent access from unauthorized sources outside the perimeter using compromised credentials. This is another variation of the same theme.

Fabulous! At this point, we have most of the major components we are going to need to create a well-planned network. Now, let's put them all together. As we do, we'll fill in a few missing bits.

Step 7 – building and configuring our foundational VPC network

Whew – that's a lot of networking bits! At this point, we have most of what we will need to implement the seventh step in our 10-step Google Cloud foundation process – build and configure our foundational VPC network – but we need to come up with a plan. It's like we have this nice, big box of network Legos; we just need to figure out what we want to build. If you check the seventh step (`https://cloud.google.com/docs/enterprise/setup-checklist#checklist-heading-7`), Google says that there are variations, depending on what you are trying to accomplish in Google Cloud, but in general, the five key network design steps we need to accomplish are as follows:

1. Set up our Shared VPC networks.
2. Decide how we want to connect our on-premises environment to our GCP environment, and implement it.
3. Configure a path for egress traffic to leave our VPC.
4. Implement security controls, including firewalls and Service Controls.
5. Decide how we will ingress and load balance traffic.

As a reminder, if you'd like to examine Google's best practices and reference architectures for VPC design, give this page a read: `https://cloud.google.com/architecture/best-practices-vpc-design`. We will mention and implement many of the suggestions as we move forward. You should also take a look at the networking section of the Google Cloud security foundations guide: `https://services.google.com/fh/files/misc/google-cloud-security-foundations-guide.pdf`.

To get started, let's enumerate the key decisions you need to make before moving forward.

Making key networking decisions

As you move through the rest of this chapter, you're going to want to make several key decisions related to your organization's network and its design. If I were you, I would copy the following itemized list to a new document somewhere. This will allow you to jot your ideas down and come up with a solid plan before attempting to implement your network.

The following is a list of key design-related questions. Details on each will be explored as we move through the rest of this chapter:

- Environment-specific Shared VPC networks? Will a hub be used?

- How many Google Cloud regions are going to be required? Which ones specifically?

- What do you know about the IP requirements for each region? Pay particular attention to heavy IP users such as GKE. You will need to create a separate document and detail your **IP Address Management** (**IPAM**) plan.

- Also, if you are leveraging some sort of hybrid cloud, remember that the IP ranges you select cannot overlap with those being used on-premises in the connected network. Make sure that you document those IPs.

- Does each environment have a homogeneous level of security, or would it be better to create a low-trust network, and then a high-trust network where you add VPC Service Controls? Do you need other isolated networks?

- What's your hybrid cloud plan? How are you going to be connecting from on-premises to Google Cloud (internet, VPN, Interconnect, or a mix)? Where will the connections be made, both in terms of the GCP region and in terms of the VPC network? This plan will vary, depending on if you are going hub-and-spoke or not.

- Will any of the VPCs need connectivity to each other? If so, and on a connection-by-connection basis, how will you implement this (peering, VPN, multi-NIC instances)?

- How and where will you build your firewalls and VPC Service Control boundaries?

- How will you ingress the access to workloads in GCP that aren't coming through the hybrid cloud connection? Will it require a load balancer? Which type?

- How will workloads in GCP egress to the outside world (external IPs, NAT)?

Before we go any further, let's talk about naming.

Updating your naming document

Do you remember our naming document? Time to add some names to it. These will apply to most environments. If you are using Cloud Routers, custom routes, Interconnect, and so on, then make sure you add names for those as well. Remember, my naming convention is based on the one Google came up with in their *Google Cloud security foundations guide* at `https://services.google.com/fh/files/misc/google-cloud-security-foundations-guide.pdf`, so look there if you'd like a fairly complete list that's all in one location:

Resource	Naming Convention
VPC	`vpc-env-code-vpc_type{-label}` Example: `vpc-p-shared-base`
Subnet	`sb-vpc_name-region{-label}` Example: `sb-p-shared-base-us-central1-net13`
Firewall	`fw-vpc_name-priority-direction-action-src_label-dest_label-protocol-port{-label}` Example: `fw-p-shared-base-1000-i-a-all-all-tcp-80`

In the preceding table, when it comes to `vpc_type`, there are four typical options:

- **service**: Designed Shared VPCs with on-premises access requirements
- **float**: Shared VPCs with no on-premises access
- **nic**: Connected to other networks and/or on-premises through a multi-NIC appliance layer
- **peer**: Networks peered to other VPCs in GCP with no on-premises access

In the preceding table, when it comes to the firewall rule names, we have the following options:

- **priority**: A number (0-65,535)
- **direction**: I(ngress) or e(gress)
- **action**: A(llow) or d(eny)
- **src-label**: All, source IR range, or tags
- **dest-label**: All, tags, or service account

- **protocol**: All or a single or combination of protocols (TCP, TCP UDP, and so on)
- **port**: The port number or range

Excellent! Now, let's start building our networking, starting with the Shared VPCs.

Planning the Shared VPCs

Before creating any Shared VPCs, you'll need to decide what the boundaries of your networks will need to be. In my current organizational design, with the different dev, non-prod, prod, and common environments, two related approaches would likely make a lot of sense. First, I create a shared VPC for each of the environments. From an isolation perspective, limiting something in dev from accessing something in prod is the highest default isolation level. The downside? It might be too isolated for some requirements. Also, where will you make the connection to the on-premises resources? A connection per environment would be the logical option, and that might be one way for the different environments to interact, but that sort of interaction may also add a lot of unnecessary latency and networking costs, since the packets would leave GCP (network egress), pass through the on-premises network, and then back up into GCP.

A variation of this approach would be hub-and-spoke. Hub-and-spoke starts with the aforementioned Shared VPC design, with spoke Shared VPC networks being created in each of the dev, non-prod, prod, and common environments. Then, it adds a hub VPC that is peered with each of the spokes. The hub is then used to bridge to the on-premises environment by exporting the routes to the spokes, allowing them to have direct communication with the on-premises network. Transitive peering limits, however, would mean that, by default, the various environments wouldn't be able to communicate with each other. Selectively peering between the spokes themselves could overcome that issue, if needed.

Now that we have an idea of a holistic design, let's talk about regions and IPAM.

Selecting Google Cloud regions and planning IPs

Before you can implement whichever VPC design will work best for your organization, you need to give some thought to the Google Cloud regions where your workloads will physically live. Remember, regions are data centers, and Google has a lot of them. Selecting regions should be tied to several key factors, as follows:

- **Proximity**: The closer (in terms of latency) you and/or your client(s) are to the Google Cloud regions you are using, the lower the latency will be when interacting with them. To check the latency from a particular location, go to http://www.gcping.com/.

If you are servicing larger areas, remember that Google has traffic routing load balancers that can easily connect clients to the region you have selected that is closest to them. I'll talk about those options shortly.

- **Regulatory compliance**: Are you dealing with any regulations that require data to be kept and/or processed within a particular geographic area? That may limit the regions you can use.

- **Feature**: Not every GCP feature is available in every region, so if there's something you are planning on using, make sure it is available in the regions you choose: `https://cloud.google.com/about/locations`.

- **Cost**: Not every GCP feature costs the same in each region, so eyeball the costs so that you're not surprised. For example, I did some work for a South American client one time who kept all their data in the US, rather than in Google's São Paulo region. When I asked about it, they informed me that they had a lot of data in Cloud Storage and that it was cheaper for them to keep it in the US, and with the way the fiber ran, the latency was still fairly low. Can't argue with that.

- **Redundancy (higher availability)**: Though the chances of a Google Cloud region or zone going down is small, it could happen. If you need systems to be up no matter what, then consider using multiple zones in multiple regions. If you already have multiple regions for proximity reasons, all the better. The only downside here is cost. Replicating systems can increase complexity, depending on what you are building, and it will certainly increase the costs.

For my design, I'm going to go with two regions: `us-central1` and `us-west1`. I'm imagining that *GCP.how* is a US-centric company, so I'll keep to regions in that part of the world. I'm using two regions because though it will cost more, I like both the higher availability and lower latency.

> **Note – Enable Private Google Access**
>
> Private Google Access is a subnet-level feature that allows resources with no public IP to access Google Cloud resources without sending traffic through Google's edge network. If you have a VPN or Interconnect to aid with your hybrid cloud configuration, then private access can be extended to on-premises workloads, but you will need to configure on-premises DNS to route all `private.googleapis.com` and `restricted.googleapis.com` traffic to `199.36.153.8/30`. Though that's a public IP, it's not announced over the internet, so the on-premises traffic will route the request over your hybrid cloud connection.

OK – with my regions sorted, it's time to put my IPAM hat on and think about IP addresses. Google recommends fewer, larger subnets, rather than smaller, application- or use-specific ones, and I agree. The larger subnets are easier to manage, and since firewalls and such can be tied to tags and service accounts and not just to CIDR ranges, I don't need the ranges as the only way to filter access. Also, don't forget the 300 subnets per VPC limit, so fewer subnets will keep us a long way from that.

The next IPAM question is, how big do the ranges need to be? Make sure that you pay attention to IP hogs such as GKE. When working with Kubernetes, Google creates a pair of secondary IP address CIDR ranges on the subnet. The secondary ranges are alias ranges – that is, they represent the IPs for containers and services running inside the GKE cluster, not the IPs of the primary machines that make up the cluster itself. You can either use normal RFC 1918 ranges for your primary and secondary ranges, or you can use RFC 1918 for the primary, and then use RFC 6598 spaces for the secondary ranges. For a full list of ranges that are supported in GCP, go to `https://cloud.google.com/vpc/docs/subnets#valid-ranges`.

My decision is going to include the following key factors:

- I want each region to utilize a single main CIDR range that I can then subdivide into smaller ranges for each of my subnets. The subnets, in turn, need to be plenty big for anything I may need to place on them. "Plenty big" is subjective. I'm going to put a /16 CIDR into each region, then sub-divide it into /18s. That will give each of my subnets 16,000+ IPs, and I'll have unallocated ranges I could use later if I decide to add subnets or regions.

- I do a lot of work with GKE, so I want to add secondary ranges, and I like the idea of using the RFC 6598s to do it. If you go with that approach, however, make sure on-premises topologies can support all the ranges you use. Also, make sure you have enough IPs in the range for whatever you are trying to accomplish. I'm going to stick to the /18 individual ranges, but if you start getting tight on IPs, remember that you can bring unused RFC 1918 ranges into the mix since the VPC doesn't have to stick to a single network-level block.

- Also, I think it makes a lot of sense to create a base network for lower security workloads and a restricted network for the higher ones. So, I'm going to need ranges to support both network types, across all my environments and regions.

- Lastly, I'm not going to leverage an on-premises connection in my example use case, so I don't have to play nicely with any external subnets. Remember, though, when you're connecting networks with peering or to external networks with VPNs or Interconnect, the IP CIDR ranges that are being used across all the linked environments must be unique and can't overlap.

With that in mind, I searched around the net and found a network IP planning tool that can lend a hand: `https://network00.com/NetworkTools/IPv4AddressPlanner/`. Playing with the tool a bit, I plugged in the base range (10.0.0.0/8, in my example) and let it know how many subnets I needed, and what size they needed to be (/18, in my example), and the tool kicked out a list of ranges I could use. I took the ranges and dropped them all into a spreadsheet where I'm currently keeping my plan. Then, I did the same thing for my GKE secondary ranges. After I was done, it looked like this:

VPC	Region	CIDR Type	Common (Hub)	Dev	Non-prod	Prod
Base	us-central1	Subnet Main	10.0.0.0/18	10.0.64.0/18	10.0.128.0/18	10.0.192.0/18
		Pod Range		100.64.0.0/18	100.64.64.0/18	100.64.128.0/18
		Service Range		100.64.192.0/18	100.65.0.0/18	100.65.64.0/18
Restricted		Subnet Main	10.1.0.0/18	10.1.64.0/18	10.1.128.0/18	10.1.192.0/18
		Pod Range		100.65.128.0/18	100.65.192.0/18	100.66.0.0/18
		Service Range		100.66.64.0/18	100.66.128.0/18	100.66.192.0/18
Base	us-west1	Subnet Main	10.2.0.0/18	10.2.64.0/18	10.2.128.0/18	10.2.192.0/18
		Pod Range		100.67.0.0/18	100.67.64.0/18	100.67.128.0/18
		Service Range		100.67.192.0/18	100.68.0.0/18	100.68.64.0/18
Restricted		Subnet Main	10.3.0.0/18	10.3.64.0/18	10.3.128.0/18	10.3.192.0/18
		Pod Range		100.68.128.0/18	100.68.192.0/18	100.69.0.0/18
		Service Range		100.69.64.0/18	100.69.128.0/18	100.69.192.0/18

Figure 6.4 – IPAM

With a base IP plan in place, let's think hybrid cloud for a moment.

Connecting the on-premises network to Google Cloud

Earlier in this chapter, we talked about our major approaches to connecting our enterprise environments to Google Cloud. As a reminder, they are as follows:

- **Internet**: Simply use your standard internet connection to reach the services in Google Cloud, exposing services using public IPs to allow access.

- **HA VPN**: This is still running through your internet, but now using a VPN to allow for private IP access.

- **Interconnect**: Dedicated or through a provider, this is typically a specialized higher bandwidth connection to Google Cloud.

When you decide on your specific VPC topology, hopefully, you give some thought to how you will connect to it from your enterprise. In my example, I'm going simple and will just use the internet and public IPs. In general, if you are going to be creating a hybrid cloud architecture, then I would recommend using the hub-and-spoke VPC design so that all the connections to your on-premises environment can be centralized.

The simplest starter plan for building a hybrid cloud framework would be to initially configure a HA VPN and grow from that into Interconnect where and when it makes sense. For a nice discussion of VPNs, Interconnects, and other network connectivity products, go to `https://cloud.google.com/network-connectivity/docs/how-to/choose-product`.

HA VPN and Interconnect options will require you to configure Cloud Routers to handle the BGP dynamic subnet sharing.

The following are recommendations for HA-VPN:

- You will need at least two tunnels from your VPN gateway in Google Cloud to your peered gateway on-premises, but more tunnel pairs are possible.

- Each active tunnel will add about 1.5 Gbps up, and 1.5G bps down, of bandwidth.

- If you have a single VPN gateway, then configure the tunnel pair in active/passive mode. If you have multiple gateways, then use active/active mode since passive tunnels don't run traffic unless the actives go down. Multiple gateways can achieve HA through multiple tunnel pairs, hence the active/active recommendation.

- Configure a BGP session per tunnel.

- Use strong pre-shared keys (`https://cloud.google.com/network-connectivity/docs/vpn/how-to/generating-pre-shared-key`).

More details about HA VPNs can be found at `https://cloud.google.com/network-connectivity/docs/vpn/concepts/overview`.

If you are going to be using Interconnect, keep the following points in mind:

- It's possible to share Interconnect connections with multiple VPCs.

- You can have up to eight 10 Gbps and two 100 Gbps connections per project.

- Ideally, anchor your Interconnect connections in two or more regions, with two or more connections each, all with an active/active configuration.

- Provide a Cloud Router to each of your connections, setting them all to global dynamic routing.

- Remember that when you're selecting your on-premises data centers, Interconnect will need to use one of Google's colocation facilities: `https://cloud.google.com/network-connectivity/docs/interconnect/concepts/choosing-colocation-facilities`.

More details about Interconnect can be found at `https://cloud.google.com/network-connectivity/docs/interconnect/concepts/dedicated-overview`.

With our enterprise to Google Cloud connectivity established, let's discuss security.

Firewalls and security

The next thing you need to plan is the firewalls you are going to need, starting with the hierarchical rules. As we've already discussed, hierarchical firewall policies let you define base firewalls at the org or folder level, allowing a high level of control over the most important (and generally applicable) rules. The goal of hierarchical rules is to set a baseline for the org and individual environments or business units. You should document your hierarchical and network-level firewalls, as well as their names, in a spreadsheet before you get to Terraforming them.

Things that may make good hierarchical policies are as follows:

- Allow load balancer health checks. They will be ingressing from ranges `35.191.0.0/16` and `130.211.0.0/22`, and `209.85.152.0/22` for ports `80` and `443`.

- Allow **Identity Aware Proxy (IAP)** TCP forwarding, the recommended way to allow SSH and RDP access to VMs, by allowing `35.235.240.0/20` for TCP ports `22` and `3389`.

- Allow Windows activation, outbound to `35.190.247.13/32` on port `1688`.

- Allow all your machines that are using RFC 1918 IPs to talk to each other, but add a `goto_next` action to delegate overridability to lower-level firewalls so that if a particular network or subnet needs to disable the connectivity between VPCed resources, they can.

Once you have a plan for the high-level hierarchical rules, start planning for the network-specific rules. Remember, firewall rules can be tied to IP address ranges, network tags (an identifier attached to a VM or service), and resource service accounts. When documenting the rules, document the method that will be applied, as well as things such as their names and actions.

The following are things you must remember:

- The implicit allow egress and deny ingress rules.

- You can allow traffic from a resource by using the network tag attached to it (such as on a VM), by the service account credential it's running under (recommended), or by its CIDR range.

- If resources must be used via load balancers, then make sure that they, and perhaps IAP TCP forwarding, are the only allowable connections to the resources.

More details on creating hierarchical firewalls can be found at `https://cloud.google.com/vpc/docs/firewall-policies`. Details on creating standard firewalls can be found at `https://cloud.google.com/vpc/docs/firewalls`.

If you are building a high-security network, then you also need to plan your VPC Service Control boundaries. Remember, Service Controls can limit things such as what data can be egressed outside the boundary. Details on VPC Service Controls can be found here: `https://cloud.google.com/vpc-service-controls/docs/overview`.

Also, if I haven't said it already, remember that if a product doesn't need a public IP for your use case, don't give it one.

Great! Now that we have a plan for the firewalls, let's talk about workload ingress.

Controlling ingress to workloads and load balancing

Some of your workloads may need to be accessible from the outside world. If so, then you need to plan out how they will be accessed.

If you expose VM instances on public IPs, make sure that you configure any firewalls you can to limit that access. Perhaps they are exposed to the outside world, but they can only be accessed when the source IP CIDR ranges match your organization's on-premises network ranges.

If you map a custom domain to a service, such as a custom domain sitting on App Engine or Cloud Run, once again, make sure you enable any firewalls that make sense.

If you are building an application on any GCP compute technology or configuring access to Cloud Storage, then you may be standing up a load balancer in front of it. Load balancers can expose a global anycast IP or a regional IP, and then balance incoming traffic over the multiple instances you have running. In this case, make sure that only the load balancer is exposed to the outside world, and block direct access to the instances either by firewalling or removing public IPs, or both.

For decisions on load balancer choices, you must know about the product where your app is running, and then do your load balancer research here: `https://cloud.google.com/load-balancing/docs/choosing-load-balancer`.

HTTP Global Load Balancer Has Lots of Cool Integrated Security

If you don't already know, Google Cloud has a fantastic, proxy-based, layer-7, global load balancer specifical designed for load balancing HTTP(s) access to workloads running on VMs, Cloud Run, Cloud Functions, App Engine, files stored in Cloud Storage, and even on-premises backends. It comes loaded with cool features such as traffic routing based on proximity (which of my deployed regions are closest to you), as well as things such as URL patterns. It also comes with some powerful security features, including Cloud Armor. Cloud Armor supports better DDoS protection and has IP, as well as geographic, allow and deny listing, along with a host of other security features. Make sure that you check out Cloud Armor if you are going to be leveraging a global HTTP load balancer: `https://cloud.google.com/armor/docs/cloud-armor-overview`.

OK – that's enough about ingress traffic. Now, let's talk a little about egress traffic.

Configuring egress options

At first glance, egress sounds easy. You need to access something; you send out the traffic. That's true if you are working on something with a public IP, but if you're following my advice and removing all the public IPs you can, then how can you allow something with no public IP to access things with public IPs?

Here, the first question is, is the thing being accessed a Google Cloud service or a general external network location? To get access to Google Cloud resources without routing traffic to the edge network, you will need to enable private access (as we mentioned earlier in this chapter) at the subnet level.

If the thing that's being accessed is external to Google Cloud, then consider enabling Cloud NAT. Cloud NAT is a fully managed, highly available **network address translation (NAT)** service. Essentially, it can loan your private IP instances a public IP when needed. For more details about Cloud NAT, go to `https://cloud.google.com/nat/docs/overview`.

Fantabulous! At this point, most of our plan should be well along the way. Now, it's time to talk about Terraform.

Terraforming your Google Cloud network

To Terraform your network in Google Cloud, your first resource should be the Terraform blueprints that Google has created at `https://cloud.google.com/docs/terraform/blueprints/terraform-blueprints`. The following blueprints can help you create your foundation network:

- `cloud-dns`: Creates and manages Cloud DNS zones and records.
- `cloud-nat`: Creates and manages Cloud NAT.
- `cloud-router`: Creates and manages Cloud Routers.
- `lb` and `lb-http`: Create and manage GCP load balancers.
- `network`: Ah, this is the big one. It can help with creating and managing VPCs, subnets, routes, peering, and firewalls.
- `vpc-service-controls`: As its name suggests.

In addition, you should look at the general Terraform resources for network-related Google Cloud resources on the main Terraform site: `https://registry.terraform.io/providers/hashicorp/google/latest/docs`.

In my example organization, I'm going to be using the network part of the example organization Google has created, as I've done in the previous chapters. I'm not going to walk you through the steps of using the CI/CD pipeline to apply the changes as we've done that a couple of times in a couple of different chapters now, and the process doesn't change. The steps are also documented in the example itself: `https://github.com/terraform-google-modules/terraform-example-foundation/tree/master/3-networks`. Pay attention to how you would configure your various hybrid cloud connectivity options. I'm not going to use anything but the internet, so I'm good to go there.

This time around, there are three Terraform Variables files:

- `common.auto.tfvars`: You'll need your organization ID and your Terraform service account (see your notes file). This is also where you can enable hub-and-spoke and transitivity if needed.
- `shared.auto.tfvars`: Here, you can set the list of DNS servers you'd like to forward requests to. This will be used to enable forwarding to DNS servers in your organization when something is encountered in GCP that needs to be routed out through your hybrid connection to something in the on-premises network.

- `access_context.auto.tfvars`: You'll have to provide the access context manager policy ID here. You can retrieve the name, if you don't have it in your notes file, using the following command:

```
gcloud access-context-manager policies list \
--organization <org_id>
```

Update your variable files and follow the instructions to push the changes through your CI/CD pipeline, as we've done previously. You can take a tour through the example organization Terraform scripts for the network if you wish. There are scripts for building the networks, the hierarchical and standard firewalls, and a DNS hub, as well as for setting up hybrid cloud connectivity. It's a nice read.

Pictures Can Help

Whether you're planning your network or all of your GCP foundation, sketching things out can help. If you don't know, Google has icons you can use – the same icons I use in many of my graphics. For Google Cloud-specific icons, see `https://cloud.google.com/icons`. For general Googly icons, see `https://fonts.google.com/icons`. For Google's new design tool (pics only, no implementation), see `https://googlecloudcheatsheet.withgoogle.com/architecture`.

And with that, it's time to wrap our network discussion and see what our Google Cloud foundation has in store for us next.

Summary

In this chapter, we continued to build our secure and extensible foundation in Google Cloud by completing step 7, where we laid out the network layer of our Google Cloud foundation. We learned how the key pieces and parts of networking worked in the context of Google Cloud. We planned our network from the top down, and our plan included everything from IP addresses to firewall names. We learned how the network helps add a layer of security to our defense-in-depth approach and how we could integrate our network with our on-premises environment. Finally, we saw how, once again, Terraform can help us create it all.

Another tough but fantastic job!

If you want to keep going through Google's 10-step checklist with me, your tutor by your side, then flip the page to the next chapter, where you will learn how to enable Google Cloud's monitoring and logging features.

7
Foundational Monitoring and Logging

In this chapter, I am going to try really, really hard not to get up on my soapbox.

Really, hard…

But seriously! How can you use Google Cloud and not know how logging and monitoring work? I mean, how can you find what's broken, know when you might be running short on capacity, or tell when something untoward has occurred if you don't have good data? And where do you get said data? Yes, Google Cloud Logging and Cloud Monitoring.

Look, I'm old enough and comfortable enough in my own skin to know the things I'm good at and the things I'm not. Don't ask me to draw you a picture or how to spell fungible, because drawing and spelling are ~~definatly~~ definitely (thank you, spell checker!) not in my skillset. But I got my first computer and started writing code in 1981, and it took me about 2 weeks to figure out that me and computers are simpático.

Figuring out what's wrong takes information. Ensuring capacity in Google Cloud, spotting bad behavior, fixing broken stuff, and, generally, knowing what's happening is all carried out thanks to monitoring and logging. So, without further ado, let's get to building our logging and monitoring foundation.

In this chapter, we are going to get to know the various tools in Google Cloud's operations suite. Then, we are going to apply them to our foundation:

- Getting to know the six core instrumentation products in Google Cloud
- Step 8 – setting up foundational Cloud Logging and Cloud Monitoring

Getting to know the six core instrumentation products in Google Cloud

In the initial stages of Google Cloud, there was App Engine, some storage offerings, and only the most basic support for logging and monitoring. As GCP grew, Google added more capabilities to Cloud Logging and Cloud Monitoring, but the available services were still lacking. So, in 2014, Google being Google bought something: Stackdriver. If you've been working with Google Cloud for a while, you might remember when the heading at the top of the instrumentation section in the Google Cloud menu was titled "Stackdriver," and you will still see references to the name in examples and documentation online to this day. In 2020, Google finally decided that the name Stackdriver should go away, and so they went to work on a replacement. The interim name they came up with was "Operations." However, while they were working on coming up with a new name with more pizazz, Covid-19 hit and the importance of getting a good name moved to the back burner.

As I sit here writing this, I wonder whether Google has just decided to stick with Operations, or whether I'm going to wake up one day soon (likely the day after this book drops) to a new group heading. Regardless, at this point, the name is Operations, and that's the name I'm using.

When you look at the current Google Cloud Console navigation menu, the subsection related to our topic at hand looks like the following:

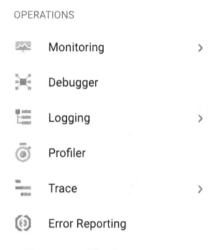

Figure 7.1 – The Operations suite

Inside the **OPERATIONS** product grouping, you will see six products. The most generally applicable instrumentation offerings are **Monitoring** and **Logging**. The other four products, **Error Reporting**, **Debugger**, **Trace**, and **Profiler**, are all designed to varying degrees to help troubleshoot misbehaving applications and to help developers.

To summarize the importance of each product, let's take a quick operations suite tour.

Instrumentation product overview

Let's start with the focus of this chapter, that is, the central two products in Google Cloud's operations suite:

- **Cloud Monitoring**: This starts with a metric, say the CPU utilization for a Compute Engine VM. The metric reports measurements of the current CPU utilization to Cloud Monitoring based on a frequency (such as every few seconds). Then, Google aggregates those measurements using math (for example, average, sum, and more) to simplify them into time-interval buckets. In the end, you have something such as CPU Utilization avg/min. Plot the measurements on a chart, and you have the makings of a monitoring dashboard. Draw a line on your chart such that below the line is good, and above the line is bad, and you have the basis of an automated alert.

Monitoring is your central tool for getting information on how things are performing now and how they perform over time. Additionally, it will feed you the data you need for various types of troubleshooting, capacity planning, and performance evaluation, just to name a few.

- **Cloud Logging**: Logging is logging. Google Cloud Logging breaks logs into several major categories. Platform logs are specific to individual services, such as Cloud Run and BigQuery, and they can answer things such as *When exactly did that request get processed by Cloud Run?* User-written logs are logs generated by applications deployed into GCP, which use the logging API to manually print messages. Security logs, which are primarily composed of audit logs, capture information on who did what, where, and when in GCP, and they help answer questions such as *Who changed this permission?* Multi and hybrid cloud logs are logs ingested from on-premises, AWS, or Azure, and help you integrate external logging into your Google Cloud environment.

We are going to spend the majority of the rest of this chapter discussing logging and monitoring as they relate to building a solid GCP foundation. However, before we do, let's do a quick tour of the four remaining instrumentation products:

- **Error Reporting**: One day this week, I was working through my emails when one of my monitors flashed bright red and then black. When it returned to normal, Apple Mail was no longer running. A few moments later, an error popped up on my Mac talking about how Apple Mail threw an uncaught exception of some kind or another and would I like to report the error to Apple. That's what happens when code blows up unexpectedly on a desktop machine. But what if the same thing happened to code running in Google Cloud? How would you know?

  Error Reporting builds on Cloud Logging to help spot unhandled errors and report them to you in a centralized way. You modify your code to send out a specifically formatted log entry, or you loop in the Error Reporting API, and when things go bad, Error Reporting notices, and can send you an alert. For more information on Error Reporting, please refer to `https://cloud.google.com/error-reporting/docs`.

- **Debugger**: The first thing to know about Google Cloud Debugger is that you don't want to use Cloud Debugger. You want to do your debugging in your on-premises dev environment before you ever even get to Google Cloud. Trust me, your on-premises debugger has many more features.

The second thing to know about Cloud Debugger is that if you can't replicate the issue in your dev environment, or if debugging on-premises doesn't really work for this issue because it needs to be running in Google Cloud first, then that's when you debug with Cloud Debugger. Unlike a typical debugger, Cloud Debugger doesn't interrupt executing code, allowing you to debug in production without any negative side effects. As with a typical debugger, Cloud Debugger augments the data you have by giving you a glimpse into what's happening inside the running code. Not only will it allow you to see the internal state when a line of code executes, such as exactly what is in that x variable when line 27 passes, but it can also help you add extra logging to the code without having to edit and redeploy it. Honestly, I'm old school, and that extra-logging-without-having-to-edit-and-redeploy thing is my favorite Cloud Debugger feature. For more information, you can start here: `https://cloud.google.com/debugger/docs`.

> **Note**
>
> At the time of writing, Google announced that Cloud Debugger was officially being deprecated and that sometime in 2023, they would be replacing it with an open source tool, so keep an eye on that.

- **Trace**: Trace is a latency tool. Let's say you build a well-designed microservice-based application. A request comes into the frontend service, and the frontend hits another service, which hits another service, which hits a database, which triggers a return back to the frontend. That internal service-to-service-to-service-to-database round-trip takes 10 seconds. You decide that 10 seconds is too slow and that you'd like to shorten the latency. OK, but to do that, you're going to need to figure out where exactly in that RPC chain the time is being spent. That's what Cloud Trace can help with. When services make RPC calls to other services, Trace can keep track of the latencies at the call points. It's a cool tool when you need it. For more information on Cloud Trace, please refer to `https://cloud.google.com/trace/docs`.

- **Profiler**: Yeah, I almost never use this tool. The theory is that you have an application running in Google Cloud and you want to know how it's spending your CPU and memory. As far as it goes, it works as advertised. The problem is that its language support is limited, and in general, there are much better language-specific profilers you can use in your dev environment. The only real use case for me would be something that I couldn't profile in my dev environment, which almost never happens.

Good. Now that you've met everyone, let's take a deeper dive into monitoring and logging, and see how our foundation leverages them. Let's start with logging.

Working with Cloud Logging

One year, two days before Christmas, I was as sick as a dog. My son and I were the only ones home as the rest of the family was visiting Mickey Mouse down in Florida. I decided to take myself into the local doc-in-the-box where a terribly young new doctor was on duty. He walked in and shook my hand, looked at my son, and said, *"I'm going to wash my hands now because your dad is running a fever."*

There, that fever? It was a piece of information about the inner state of my body. The doctor used that information to start his search for exactly what was making me feel bad.

For the most part, systems running inside Google Cloud operate as black boxes. You see them, know they are there, but they tend to hide a lot of their inner workings. To a certain degree, all of Google Cloud is a black box. You might have an account, perhaps quite a large one, but that doesn't mean Google is going to let you take a tour of a data center. We've all seen that movie where James Bond or someone sneaks off the tour to hack some server, right? Google isn't having it, and since you can't just walk over and poke at a server in the data center, you are going to have to rely on the data Google does provide to get answers.

Generally, we have two choices when it comes to instrumenting systems. First, if we have full control over all aspects of a system, perhaps because we wrote the code ourselves, there might be various ways to make the box clearer and see exactly what's happening inside it. For systems we don't have control over, our only choice is to rely on the telemetry being passed out of the black box into Cloud Logging and Cloud Monitoring.

Sometimes when I teach, I have a short example application that I use as a demo: https://github.com/haggman/HelloWorldNodeJs. It's written in Node.js (JavaScript), and it's easy to run in Google Cloud's App Engine, Kubernetes Engine, or Cloud Run. It's a basic Hello World-level application running as a web application. You load it up into GCP, visit its home page, and it returns Hello World!. The source code is as follows:

```
const express = require('express');
const app = express();
const port = process.env.PORT || 8080;
const target = process.env.TARGET || 'World';

app.get('/', (req, res) => {
```

```
    console.log('Hello world received a request.');

    res.send('Hello ${target}!');
});

app.listen(port, () => {
    console.log('Hello world listening on port', port);
});
```

The two lines I want to draw your attention to are the ones that start with console.log. In JavaScript/Node.js applications, console.log is a way to log a message to standard-out. That's programming speak for the default console location. Normally, that would be the command line, but with code running on some server, it typically ends up in some log file. The second console.log line prints a message to let you know the Express web server is up and ready to work. When you visit the application's home (/) route, the first console.log line logs a message just before the app prints **Hello World!** to the browser.

If I were to run the preceding application from the Google Cloud Shell terminal window, then the log messages would appear on the command line itself, since under those conditions, standard-out would be the command line. However, if I deploy the application to Cloud Run and run it there, the log messages will get redirected to Cloud Logging.

> **Best Practice: Log to Standard-Out**
>
> Traditionally, applications manage their own log files. You create a my-app/ logs folder or something, and manually write your logs into it. In Google Cloud, that's not a good design, even for applications designed to run on their own VMs. As a best practice, code your application so that it logs to standard-out, and then let the environment determine how best to handle the messages. If you need a little more control, for example, you want Cloud Logging to create a special log just for your application logs, then use the Cloud Logging API or a logging utility that understands how to work with Google Cloud. For example, in Node.js, I can easily use Winston or Bunyan since they both have nice integrations for Google Cloud Logging. For other examples, please refer to https://cloud.google.com/logging/docs/setup.

For demonstration purposes, I just cloned my Hello World code into Cloud Shell. I built the container and loaded it into Google Container Registry using Google Cloud Build. Yes, that's the same Cloud Build I'm using to handle my infrastructure CI/CD pipeline. Then, I deployed the container into Cloud Run. When I visit the Cloud Run service test URL, I see my *Hello World!* message.

Before I discuss specific logs, let's take a quick detour and check out Google Cloud Logging's Logs Explorer. The **Logs Explorer** option can be found in the **Logging** section of the navigation menu. It's Google's central tool for viewing logs. As of the time of writing, it looks similar to the following screenshot:

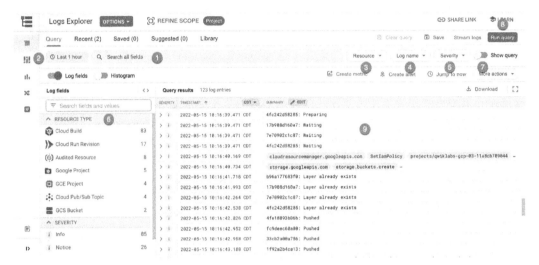

Figure 7.2 – Logs Explorer

In the **Logs Explorer** window, you can see the following:

1. Insert something you'd like to search the logs for.

2. Set a window of time for the search; the default setting is the last hour.

3. If you prefer, you can search by resource type.

4. Or you can search by log name.

5. Or you can search by log severity.

6. Even easier, you can simply apply a filter. The filters on this list are context-sensitive, in that they change depending on the log values being returned. Essentially it's Google's way of saying, *"In the logs I'm seeing, these might be some ways you'd want to filter them."*

7. If you prefer, you can display the log query and edit it directly before you run it.

8. When you are ready, run the query.

9. The results will be displayed here. In this case, you are looking at the middle of a set of logs coming out of the Cloud Build job that built my demo application.

Speaking of my demo application, what kinds of log messages do you think were generated when I tested it? The answer is a bunch of them.

Google breaks logs into categories. A good discussion of the available logs in Google Cloud can be found at `https://cloud.google.com/logging/docs/view/available-logs`. I want to focus on the main three: platform logs, user-written logs, and security logs.

First, there are the platform logs. Platform logs are generated by specific Google Cloud products. More information on platform logs can be found at `https://cloud.google.com/logging/docs/api/platform-logs`. In the preceding screenshot, you are seeing logs coming out of the Cloud Build job to build my application's container. If I wanted, I could use the log fields filter (#6 in the preceding screenshot) and filter out everything but the Cloud Build logs. Instead, I'm going to use the log name filter (#4) and filter the entries to the Cloud Run platform logs. The results look like this:

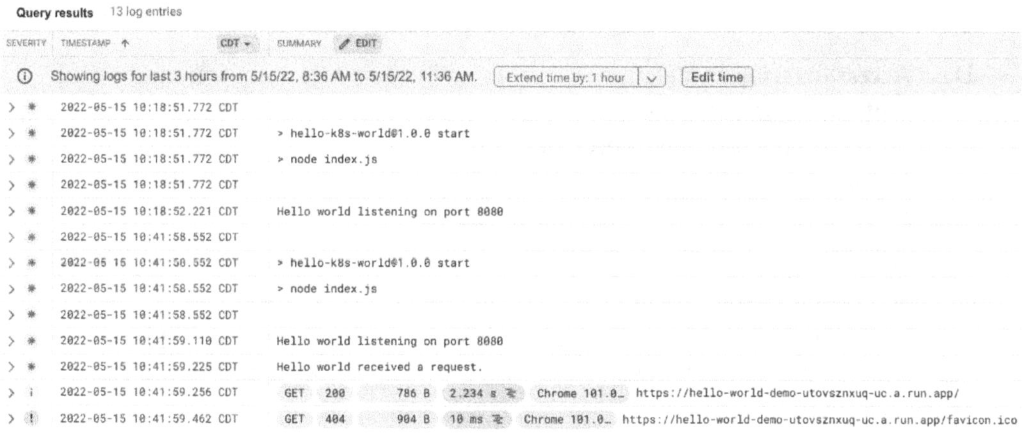

Figure 7.3 – The Cloud Run logs

The bottom two entries in the preceding screenshot show my test HTTP GET request, which was successfully handled (200), along with my browser automatically asking for the page's favicon.ico. I haven't set this up, hence the unhappy 404-not-found response. Above those messages, you see the logs I'm generating in my code. There are two Hello world listening on port 8080 messages indicating that Cloud Run must have fired off two containers, each one indicating that it's ready to receive traffic. Also, I can see a single message when I requested the application's home page: Hello world received a request. Since I'm logging to standard-out in my code, Cloud Run captures these messages on my behalf and loads them into Cloud Logging.

Another log message type would be user-written logs. What I did in the preceding Hello World example isn't technically considered user-written (pedantic much?) because I didn't create my own log file to do it. If I had instead used the Cloud Logging API, such as in the example at `https://github.com/googleapis/nodejs-logging/blob/main/samples/quickstart.js`, to create a log named `my-log` and written to that, then it would be classified as user-written. Yes, it's hair-splitting, but for now, let's stick to Google's definitions.

The last log type I'd like to mention, and arguably the most important to laying a good Google Cloud foundation, is security logs. The majority of your security logs are contained within what Google calls Audit Logs (`https://cloud.google.com/logging/docs/audit`). They include the following:

- **Admin Activity logs**: These are changes to resource configurations or metadata. Always on, always free, immutable, and stored for 400 days, this is where you look if you want to answer questions such as who added that new VM? Who changed that permission? And where did that Cloud Storage bucket come from?

- **Data Access logs**: Off by default for everything but BigQuery, not free, and stored, by default, for 30 days when enabled, these logs contain entries indicating user access and the change of data stored in Google Cloud. Since these logs are mostly off by default, you are going to have to decide whether you want to enable these logs for more than just BigQuery or not. More to come.

- **System Event logs**: These are changes made to your resources by automated Google systems. For example, if Google needs to upgrade the hardware where your VM is running and moves it seamlessly to a new box, that's a log event in the system event logs. That example is called a live migration if you are Google Compute Engine curious.

- **Policy Denied logs**: These are generated when a user attempts to do something in Google Cloud but are prevented by a VPC Service Controls security policy. So, a log entry is generated when you tried to access the files you had in Cloud Storage but are denied, not because you don't have access, but because you are attempting to access the data from your home machine, which is not within the configured service perimeter.

- **Access Transparency logs**: These can be enabled if you have the right type of support package. Once enabled, if you open a ticket with Google Support, and the support rep needs to access some of your data or systems while troubleshooting, then these logs will be generated. In addition, you can enable access approval, and the support rep will have to check with you for approval before touching your resources.

> **Note**
>
> This log will not generate entries if Google happens to touch something of yours while troubleshooting a lower-level problem. So, if Cloud Storage is broken for multiple clients, and they look at your files while troubleshooting, there will be no log entry.

To see an example of a security log at work, let's look at the admin activity log. In this example, I've gone into Logs Explorer and used the log name filter to restrict to the **Cloud Audit | Activity** log. Additionally, I've used the log fields filter to only show the Cloud Run Revision entries. So, I'm asking Logs Explorer to just show me the Cloud Run-related admin activity log entries. The results are displayed as follows:

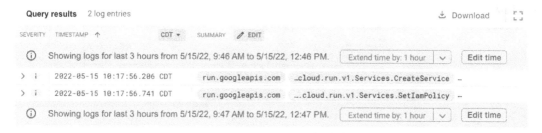

Figure 7.4 – The Cloud Run admin activity logs

The two entries are where I first created the service for my Hello World demo application, and where I set its permissions to allow for anonymous access. Nice.

Before I wrap up the discussion of logging, for now, I'd like to talk a little about log exports.

Configuring log entry destinations

By default, all your logging information is collected at the project level and stored in log storage buckets. The flow of the logs resembles the following diagram:

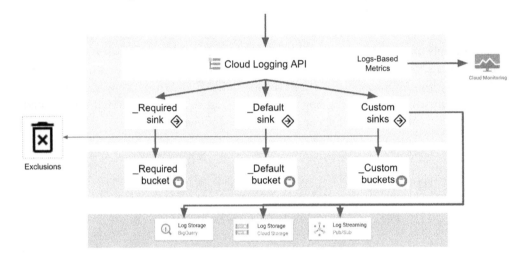

Figure 7.5 – Routing logs

So, the steps logged messages run through are similar to the following:

1. Log entries are generated, and they enter through the Cloud Logging API.

2. If a logs-based metric has been created, then matching entries are tallied and reported as measurements to Cloud Monitoring.

3. All log entries are then passed to each configured log sink. The sinks then identify the entries they are interested in and forward them to a specific destination.

The log sinks might be configured with exclusion and inclusion rules. The exclusions are processed first, ignoring any log entries matching a pattern. Then, the inclusion filter decides which of the remaining log entries should be captured and replicated into the configured destination.

Log entries can be routed into the current project or a remote one, where they will be stored in one of the following resources:

*   **Log buckets**: These are specialized Cloud Storage buckets designed to store logging entries. This is the only sink destination that is queried by Logs Explorer in Google Cloud Console.

*   **BigQuery datasets**: Exporting log entries to BigQuery allows finer-grained queries from a wider audience of data analysts.

- **Cloud Storage**: This is, typically, used for inexpensive long-term storage, possibly with retention and/or auto-delete life cycle rules.

- **Pub/Sub**: This is a good option when you need to route entries for secondary processing using some sort of code. The code could be custom created or it could be a third-party log management tool such as Splunk.

The two initial log storage buckets are `_Default` and `_Required`. `_Required` is a non-configurable storage bucket fed by a non-configurable log sink. The admin activity, system event, and access transparency (if enabled) logs are routed here and stored for 400 days. These are your core security audit logs, so when you are trying to figure out how that resource got removed or who modified that permission, you can find the details here.

The `_Default` log bucket and corresponding sink capture all the other log entries and store them for 30 days. Both are configurable.

> **Note: Not All Logs Are Enabled by Default**
>
> One of the key decisions you will need to make when configuring your logging foundation is which optional logs to enable. Generally, more information is better when you are trying to troubleshoot something. Having said that, logs, as with most things in Google Cloud, can cost extra. As such, you will need to give some serious thought to the logs you have enabled all the time, and those that you only enable when troubleshooting. We'll go into more detail on that soon.

Now that we've had a taste of Google Cloud Logging, let's head over and see what Cloud Monitoring has to offer. Then, we can get to laying that foundation.

Monitoring your resources

Have you ever read Google's books on building reliable systems? If not, head over to `https://sre.google/books/` and take a look. At the time of writing, Google has three **Sight Reliability Engineering** (**SRE**)-related books, all of which you can read, in full, from the preceding link.

In Google's original SRE book, they have a diagram that resembles the following:

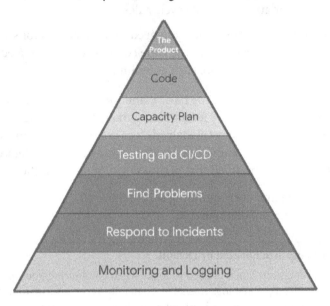

Figure 7.6 – The SRE pyramid

At the top of the pyramid is the application you are building, the product. To get the product, first, we must do the work and write the code. However, if our application is going to work well, then we need a capacity plan, and we need to follow good testing and CI/CD release procedures. At some point, something is going to happen, and we are going to need to be able to find and fix problems in response to a formalized incident response. All of that will depend on us having good information, and that information will come from monitoring and logging.

We've discussed logging, so how does monitoring fit in?

Monitoring helps us to understand how well our application is doing. Do you remember my story about being sick and missing Disney World? When the doctor takes a measurement of your temperature, that's a form of monitoring.

Have you ever heard of the four golden signals? In the same original SRE text that I mentioned earlier, Google lays out what they call the four golden signals. The gist is KISS. When you start monitoring, begin with the four most commonly important pieces of performance information, and then expand from there.

The four golden signals are listed as follows:

- **Latency**: How long does it take the application to get an answer back to the caller? Latency is frequently one of your non-functional requirements.

- **Traffic**: How much load is on the system? Typically, this is measured as requests per second, queries per second, bits per second, and the like.

- **Errors**: How many errors are being generated in a particular window of time?

- **Saturation**: The least obvious of the golden four, this is a measure of fullness. Something in your system will have a maximum capacity or load that it can handle. If your database supports 8,000 QPS, and your current load is 6,000 QPS (a traffic measure), then you are sitting at 3/4s saturation.

The four golden signals might be conceptually obvious, but how do you collect the data you need for each? Ah, that's where monitoring comes in.

To monitor products in Google Cloud, my advice is for you to work your way through the following three steps until you find what you're looking for:

1. Check the Google Cloud resource you want to monitor and see whether it has any built-in monitoring. If it does, and most Google Cloud products will, Google is offering you an opinionated view of important things to pay attention to. Many times, you won't have to go beyond this step.

2. Check to see whether Google has created a dashboard for what you want to monitor in **Monitoring | Dashboards**. If so, then this is another place where Google is attempting to show you what to pay attention to.

3. If neither of the preceding steps provides you with the information you want, then use **Monitoring | Metrics Explorer**. To use it, first, you need to spend some time in the documentation (start at `https://cloud.google.com/monitoring/api/metrics`) researching Google's available metrics.

Let's run through the monitoring checklist one at a time using my Hello World application. In the folder containing my app, there's a sub-folder that has a Python load generator that can run out of a Kubernetes cluster. I've spun up a GKE cluster and deployed the load generator into it. I have it configured to throw 30 RPS of load onto my Hello World application, which I've deployed to Cloud Run.

So, if I wanted to monitor an application running in Cloud Run, the first option says that I should start by checking to see whether Cloud Run has any monitoring. And yes, it does! Here, you can see a screenshot of the **METRICS** tab of the Cloud Run page for my application:

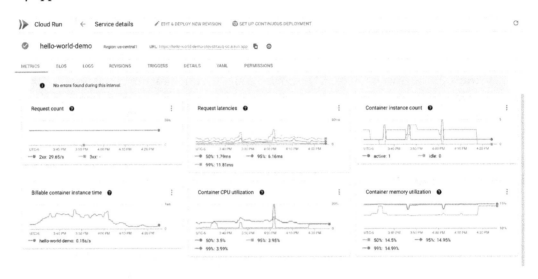

Figure 7.7 – Monitoring Cloud Run

Look at all that nice data. There's a RPS count, which shows that I'm pretty close to my 30 RPS target. There are request latencies for the 50th, 95th, and 99th centiles, which shows me that 99% of all the requests are getting an answer back in under 12 ms. The container instance count shows that there were several containers for a while, but Google has figured out that the 30 RPS load can be handled by a single active container, and that's all that's running now. On the second row, I see billable container instance time because Cloud Run bills for every 100 ms of execution time. Then, there's container CPU and memory utilization. In Cloud Run, you can configure the CPU and memory resources provided to a running container, and the setting I'm running under now, 1 CPU and 512 MiB memory, looks to be more than enough. As of the time of writing, Google has just previewed partial CPU Cloud Run capabilities, and if I could drop the CPU my service specifies, I could likely save money.

See? Monitoring.

Not bad. Usually, with Cloud Run, I don't need to look any further. If I did, I'd find that option 2's pre-created dashboards, **Monitoring | Dashboards**, don't give me anything for Cloud Run. However, there is a nice dashboard for GKE, so you might want to check that out at some point.

Moving on to option 3, Metrics Explorer, requires some prep work. Metrics Explorer has a lot of options on exactly how it can display a selected metric (`https://cloud.google.com/monitoring/charts/metrics-explorer`), so make sure you spend some time learning the interface. Before you can even start though, you need to do a bit of homework to figure out the specific metric that you'd like to examine.

Researching metrics starts at `https://cloud.google.com/monitoring/api/metrics`. For my example, I'm going to follow the link to *Google Cloud metrics*, and then scroll down to the Cloud Run section. At the top of this section is the base metric name. In this case, it's `run.googleapis.com/`, so all the Cloud Run metric's technical full names will start with that.

Next, scroll through the metrics until you find the one you need. For our example, let's go with the following:

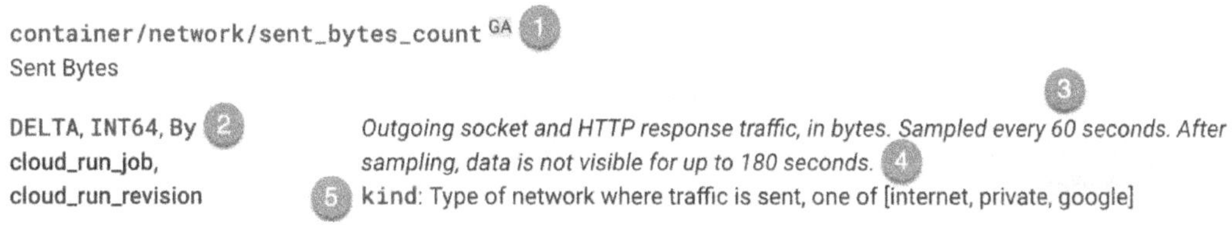

```
container/network/sent_bytes_count GA  1
Sent Bytes
                                                                           3
DELTA, INT64, By  2         Outgoing socket and HTTP response traffic, in bytes. Sampled every 60 seconds. After
cloud_run_job,              sampling, data is not visible for up to 180 seconds.  4
cloud_run_revision           5  kind: Type of network where traffic is sent, one of [internet, private, google]
```

Figure 7.8 – Cloud Run metric

So, this is a metric that will show the amount of data downloaded from your Cloud Run application, with several details. Specifically, it shows the following:

1. The technical and human-readable versions of the metric name.
2. This shows the metric's type. In this case it's a whole number (**INT64**) representing the number of bytes (**By**) of data. **DELTA** tells us that the metric is measuring change (as opposed to a gauge metric, which would show you an amount, like the gas gauge in a car).
3. A metric description, including how frequently the metric is sampled or measured.
4. Information on how long it takes after a measurement is taken before the new value shows in Metrics Explorer.
5. Any available labels, or ways to group/filter the metric.

Heading over to Metrics Explorer and searching for the metric using either the technical or human-readable name (double-check the full technical name to ensure you're using the right metric) will allow you to see the visualization. In this case, it would look similar to the following:

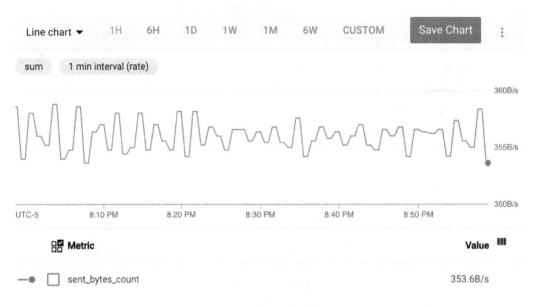

Figure 7.9 – Sent bytes

Nice. So, it's an easy view displaying the amount of data being downloaded by your application. With a little research, Metrics Explorer can be a powerful tool, but again, using options one or two first will generally save you time.

Now that you understand how monitoring works for many of the core Google Cloud resources, let's take a moment to discuss the VMs running Google Compute Engine.

Monitoring agents and Google Compute Engine

Imagine you have a VM running in Google Compute Engine. From Google's perspective, your VM is probably one of several running on the same physical server in a Google Cloud compute zone. Google can see your VM as it runs and tell how much CPU you are using, how big your disk is, and how much data you're sending across the network because all of that is happening at the physical server level. What Google can't do, by default, is see what's happening inside of your VM. Google can't see the logs being generated by that application you have running, or how much of the disk and memory is being used by said application.

For that, you need to get an inside view of the VM's black box. The traditional option would be to log into the VM using RDP or SSH and look at the logs and metrics from the OS. That's fine, and not only will it work in GCP, but it also won't add anything to your Google Cloud bill. The problem is that the VM is likely a very small part of what you are doing in Google Cloud. Since the VM isn't pushing its logs out to Cloud Logging, you can't really go to a single place and see the VM logs side by side with what's happening in that GCP data storage service that your VM is accessing. To make that work, you need to open a window into the VM that Google can see through. That window is created by Google's Ops Agent: `https://cloud.google.com/stackdriver/docs/solutions/agents/ops-agent`.

The Google VM Ops Agent can be installed in a number of ways, including manually via SSH, from the **Observability** tab in the VM's details page in Compute Engine, using the **VM Instances** dashboard, or you can create an Agent Policy (`https://cloud.google.com/stackdriver/docs/solutions/agents/ops-agent/managing-agent-policiesz`) and automate the Ops Agent installation across entire sections of your organizational resource hierarchy. Honestly, the best practice is to install the Ops Agent into the VM image itself, and then use nothing but custom images in production. If you need a little help to get that working, take a look at HashiCorp Packer.

The Ops Agent is configured using YAML and can stream system logs, application logs, and resource utilization metrics out of the VM and into Google Cloud Logging and Monitoring. Nice.

Now, before we move on to laying our logging and monitoring foundation, let's talk a little about how metric scopes work.

Metric scopes

If you visit the **Monitoring | Overview** page in Cloud Console for one of your projects, in the upper-left corner you'll see a **Metrics Scope** link, similar to the one shown in the following screenshot:

Figure 7.10 – Link to Metrics Scope

Opening your **Metrics Scope** link will display a page resembling the following:

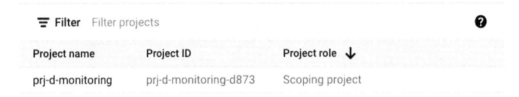

Metrics Scope   ✕

This project might be monitoring metrics from multiple other projects. The tables below
list which metrics this project is monitoring, and which projects are monitoring this
project's metrics. Learn More

Metrics monitored by this project

≡ **Filter**   Filter projects   ❷

Project name	Project ID	Project role ↓
prj-d-monitoring	prj-d-monitoring-d873	Scoping project

Add Cloud projects to metrics scope

The projects listed below can view this project's metrics

This project's metrics are visible only in this project

Figure 7.11 – Current Metrics Scope

A metrics scope is a group of one to many projects that can be monitored from the
currently viewed project. Let's imagine you have a microservice-based application with
services spread out over five different projects. Since the different projects are different
parts of the same application, sometimes, you just want to go into project 3 and monitor
the resources you have there. That's the default, and each project comes preconfigured
with a metric scope containing itself. So, if you want to monitor project 3's resources, then
just go to project 3 and do so.

But metric scopes could be edited, and you can add other projects to what's monitorable
from here. So, in my five-project microservice story, you could create a project 6
specifically to help monitor the application as a whole. You add projects 1–5 into the
metrics scope for project 6, and now you can use project 6 to monitor resources across the
entire application.

Perfect. So, we've seen how logging works and how monitoring can help. Now, let's put
them together and get them into our foundation.

Step 8 – setting up foundational Cloud Logging and Cloud Monitoring

Every resource you use in Google Cloud will impact how and what you collect in terms of logging and monitoring information. At this point, our goal is to get a few logging and monitoring fundamentals in place as part of our foundation, and then we can extend that foundation as needed. You will need to decide which of the following recommendations makes sense in your environment, and then adapt them as needed.

Let's start with logging.

Logging foundation

Let's make this clear. Logging is good because it contains information needed to spot security issues, troubleshoot applications, and more, but logging also costs money. I'd love to simply say that you should enable every type of log you can, but the reality is that the cost would likely outweigh the possible benefits. The following recommendations try to balance information needs and cost. Regardless of whether you go with my recommendations or not, make sure that you spend some time in your billing reports, so you know exactly how much you are spending as it relates to logging and monitoring. Do not get surprised by instrumentation spending.

Here is my list of logging baseline recommendations:

- **Admin Activity logs**: They are already on and free, so you're good.

- **Access Transparency logs**: If your support package includes Access Transparency logs, then turn them on. We'll talk more about support packages in *Chapter 8, Augmenting Security and Registering for Support*.

- **Ops Agent**: Install this in all of your VMs, but monitor and set budgets on the related costs. Depending on what you're running on the VMs, these logs can get big. Generally, I think they are worth the cost, but remember that you can configure both the Ops Agent and Cloud Logging sink exclusions to filter down the entries if so desired.

- **VPC Flow Logs**: Sample network packets as they move across the VPC. They are off by default, and I'd leave them off unless you are actively troubleshooting a VPC network connection problem.

- **Firewall Rules Logs**: They contain each admitted decision and rejected decision made by your firewalls. They are another set of off-by-default logs that can get really big, really fast. I'd leave them off unless you are troubleshooting firewalls.

- **HTTPS Load Balancer logs**: They are on by default. I recommend leaving them on, but you might want to exclude most (exclusions can be percentage-based) of the successful requests and log any HTTP errors.

- **Cloud Identity logs**: Check the Cloud Identity configuration options as there is a way to share its logs with Google Cloud. Enable it so that you can see all of the related changes and authentication attempts inside Google Cloud.

- **Cloud Billing**: We've already discussed exporting these logs to BigQuery for better analysis, so we should be good here.

- **Data Access logs**: They are off by default for everything but BigQuery. These logs contain any access or change to data made by one of your users, and they can be selectively enabled resource by resource and project by project. Turning them on can generate a lot of logs, but they will also allow you to answer questions such as *Who modified that file in Cloud Storage?* But did I say they could get big?

  Here, you are going to have to do some planning. If you like the idea of data access logs, then you can enable them for all projects, but you need to spot-check how much you are paying. Typically, this is where I start, by enabling them across the entire organization.

  If my data access logs are getting too big and expensive, then one possible cost-savings measure is to enable data access logs for production projects across the entire production organizational folder, but disable them in dev and non-production. If you need to troubleshoot something in dev, you can enable them temporarily. Perhaps enable them for selected resources rather than all resources. Additionally, you can use log exclusions to support enabling data access logs, but then only sample the log entries.

For some of the preceding optional logs, enabling them with Terraform will be part of setting up that particular resource. Firewall logs will be enabled in the Terraform resource for creating the firewall, and the VPC Flow Logs will be enabled when creating a subnet. If you'd like to install the Ops Agent in your VMs, you can do that when configuring the VM itself, or you might look at Google's blueprint resource to create Ops Agent policies at `https://github.com/terraform-google-modules/terraform-google-cloud-operations/tree/master/modules/agent-policy`.

Of course, for the data access logs, the easiest way to enable them is using Terraform. Enable them for the org using `google_organization_iam_audit_config` and/ or for folders using `google_folder_iam_audit_config` resources. For example, to enable data access logs for all the services within a folder, you could use the following:

```
resource "google_folder_iam_audit_config" "config" {
  folder = "folders/{folder_id}"
  service = "allServices"
  dynamic "audit_log_config" {
    for_each = ["DATA_WRITE", "DATA_READ", "ADMIN_READ"]
    content {
      log_type = audit_log_config.key
    }
  }}
```

If you are using Google's example foundation scripts that I used in earlier chapters, then you can enable the data access logs in the `1-org` step by setting the `data_access_logs_enabled` variable to `true`. To see how the script works, examine the `envs/shared/iam.tf` file.

Once you've made some decisions regarding the logs to enable, give some thought to log aggregation.

Aggregating logs

Log aggregation gives you the ability to export logs from multiple projects into a single searchable location. Usually, this will either be a BigQuery dataset or a Cloud Storage bucket. BigQuery is easier to query, but Cloud Storage could be less expensive if you are storing the logs long-term for possible auditability reasons. I prefer BigQuery with a partitioned storage table. I mean, what's the use of copying all the logs to a single location if you aren't going to be searching through them from time to time? Partitioning the table will make date-filtered queries faster, and the storage will be less expensive for partitions that are 90 days old or older.

For log aggregation, first, create a centralized project where you can store the logs. If you are using the Google example org Terraform scripts, then the project created for this is in the `fldr-common` folder and is named `prj-c-logging`.

Log aggregation can be configured with the `gcloud logging sinks` command or by using Terraform. Google has a log export module that you can find at `https://github.com/terraform-google-modules/terraform-google-log-export`. It provides facilities to create a BigQuery, Cloud Storage, or Pub/Sub destination, which is then used to sink the specified logs. Once again, if you are using the example foundation scripts, log sinks are created and configured in the `1-org` step. To see how the script works, examine the `envs/shared/log_sinks.tf` file.

With the key decisions made on how to store logs, let's talk a bit about monitoring.

Foundational monitoring

On the monitoring end, exactly what you need to view will be a project-by-project decision. In other words, if you need to monitor a particular service that you've created, do that in the service's host project. Foundationally speaking, it might be a nice idea to create a central project within each environmental folder for when you need a single pane of glass. So, in the dev folder, create a project with the metrics scope configured that can see all the other projects in the dev environment. When you need to monitor some services in `dev-project-1` and `dev-project-2` from a single location, this project will be set up for that.

If you are using the example foundations scripts from Google, the central monitoring project is created in the `2-environments` step of the `modules/env_baseline/monitoring.tf` script. Essentially, it's just a project created using Google's project factory blueprint with a name such as `prj-d-monitoring`. However, the script doesn't actually configure the metrics scope. You can either configure that manually using the UI or via the monitoring API. In the UI you would do the following:

1. Sign in to the central monitoring project for the desired environment.
2. Navigate to **Monitoring | Overview | Metrics Scope**.
3. Identify all the projects you'd like to add, and then click on **Add Cloud projects to metrics scope**.

There's also a (currently beta) Terraform resource that you can use: `google_monitoring_monitored_project`.

> **Note: A Metrics Scope Can Contain No More Than 375 Linked Projects**
>
> So, if you have more than 375 projects in your dev/non-prod/prod environment folders, then you would need to create multiple "single-pane-of-glass" monitoring projects. Perhaps you have 1,000 different service projects, but they make up 25 major application types. In that case, perhaps you will create 25 different monitoring projects, one per major application type.

Excellent! With an idea of the things that should be in your monitoring foundation, let's take a moment to point out a few other considerations.

Food for thought

I could seriously write a book on logging and monitoring in Google Cloud. In this chapter, I've tried to focus on the things that should be part of most Google Cloud foundations, without getting into all the variations you'll need to implement depending on exactly what you are doing in Google Cloud. Now, let's talk about some stuff you should at least consider.

Security Information and Event Management (**SIEM**) tools are third-party tools designed to help you scan your logs for untoward activities. Splunk, Elasticsearch, Sumo Logic, ArcSight, and QRadar, just to name a few, could all be configured to connect to Cloud Logging using Pub/Sub. Then, they can analyze what's happening in your logs and spot naughty people trying to do naughty things. Additionally, Google now has Chronicle, which can scan logs alongside information coming out of Security Command Center for various threats.

Homegrown log analysis could be constructed in a few different ways. You could centrally aggregate your key logs into a Pub/Sub topic, read the topic with Apache Beam running in Dataflow, and then dump your logs into BigQuery for longer-term storage. Then, the Dataflow/Beam code could scan the log entries for anything you liked and alert you of findings in near real time. It would require custom coding, but you could grow the solution how and where you need it. I guess it all depends on whether you're part of a build-it or buy-it sort of organization.

Another way you could do homegrown log analysis would be using BigQuery. Use Cloud Scheduler to run a job in Cloud Run or to call a cloud function once every 15 minutes. The function code executes a query in BigQuery looking for events since the last run. You could scan for admin activity log entries, logins using high-privilege accounts, IAM permission changes, log setting changes, permissions changes on important resources, and the like. If anything is found, send out a notification.

If your issues are hybrid-cloud related, and you are attempting to integrate systems running outside GCP into Cloud Logging and Monitoring, then check out Blue Medora BindPlane. It supports integration with Azure and AWS, along with on-premises VMware, and various application and database types.

Last but not least, please give some thought to how you are going to build reliable systems in Google Cloud. Read those Google SRE books I mentioned earlier (`https://sre.google/books/`) so that you can put the SRE teams in place, have good plans for your formalized incident response, and understand more about how to build your SLOs.

Cool beans. And with that, it's time to wrap up our foundational monitoring and logging discussion and move on to our final two Google Cloud foundation steps.

Summary

In this chapter, we continued to build our secure and extensible foundation in Google Cloud by completing step 8, where we implemented key parts of our logging and monitoring foundations. We overviewed the tools that Google has for us to help with instrumentation, and then we focused on logging and monitoring. We explored how Google ingests and routes logs, and we looked at the various log types. Then, we moved on to monitoring and saw the various ways Google helps us monitor the services we have running inside of Google Cloud. Lastly, we discussed the additional elements our foundation needed as it relates to logging and monitoring.

Great going, y'all! Keep on hanging in there, and we are going to get this foundation built.

If you want to keep on moving through Google's 10-step checklist with me, your personal tutor, by your side, flip the page to the next chapter where we are going to discuss our final two foundational steps: security and support.

8
Augmenting Security and Registering for Support

It's hard to believe, but we're down to our last two foundational layers: **security** and **support**. In a way, holding off on getting your support package in place until you are close to going live with production systems makes a bit of sense, but security? Zero trust security is something you have to do in layers, and that's why we've been securing our foundation one step at a time since *Chapter 2, IAM, Users, Groups, and Admin Access*. I mean, the first thing we did was to set up our identity provider. Then, we configured the users and groups, got some administrative access in place, figured out who was paying for everything and who could control billing, set up our initial resource hierarchy, configured user access, built our initial network, and set up some logging and monitoring. Think about all the security that we have in place already.

In this final chapter, our goal is to review elements of Google Cloud security that you should consider, augment our existing security components with a few final measures, and finally, set up a support plan with Google for when we need a little extra help.

In this chapter, we will cover the following topics:

- Step 9 – augmenting foundational security
- Step 10 – setting up initial Google Cloud support

Step 9 – augmenting foundational security

Security is something that you should never (and Google will never) quit working on. Technologies and the vectors used to attack them, change with time, so it's a good thing you are partnered with one of the premier IT security companies in the world: Google. Google's trusted infrastructure provides a multi-layered, zero trust approach to security that we can take advantage of when we deploy our resources into Google Cloud (`https://cloud.google.com/security/infrastructure`). Google has hundreds of full-time security personnel doing nothing but keeping things in Google Cloud safe.

Having said that, deploying your IT services into Google Cloud does not mean that all of your security is magically complete. You and Google are part of a shared fate/shared responsibility environment. Google builds and secures the data centers, the servers, and the network, and makes sure the people working in those data centers on that hardware are authorized to do so and are doing what they are supposed to do. But Google doesn't have full control over how you use their environment. To make things even more interesting, depending on the product you select, you might have a larger or smaller piece of that responsibility pie. Google does have professional services and support to help because, let's face it, if you get hacked in Google Cloud, it's going to give both you and Google black eyes, for sure. That's the shared fate in this discussion. Regardless, ultimately, it's your responsibility to know where Google's control ends and yours begins.

Let's start our security augmentation measures with a quick overview of data encryption in Google Cloud.

Data encryption

I remember an exam question that essentially read as follows: *"True or False. Strong encryption is always an option for data stored in Google Cloud."*

It sure sounds like something that should be true, right? But what's that line from Star Wars? *"It's a trap!"*

It is not true that data encryption in Google Cloud is always an option because data stored in Google Cloud is *always* encrypted, period. Any customer data stored in Google Cloud, in any GCP product from Cloud Storage to BigQuery, to persistent disks on your Compute Engine VMs, is always encrypted. Most data that is stored in Google Cloud ends up in Colossus, Google's massively scalable storage array. To understand how Colossus encrypts data, look at the following diagram:

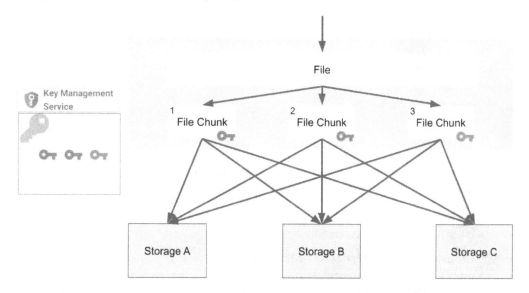

Figure 8.1 – Encrypting data in Colossus

First, Google takes your data, as shown by the file in the preceding diagram, and breaks it into multiple chunks with each chunk containing up to several GB of data. Then, Google uses a unique **data encryption key** (**DEK**) to encrypt each chunk, such that no two chunks will have the same key. Next, to achieve the 99.999999999% durability that Google's Colossus storage array provides, the chunks are replicated out over storage several times. Finally, Google uses a **key-encryption key** (**KEK**) stored in their own version of the **Key Management Service** (**KMS**), to encrypt and store all the DEKs.

That all means that if a hacker wanted to get access to your file from Cloud Storage by hacking Colossus, then they would not only have to identify and get each of the data chunks that comprise your file, but they would then have to break the encryption chunk by chunk. To make things even more exciting, the physical storage is AES encrypted, so even if the hacker was an insider with physical access to the storage, they would have to hack each drive before getting to the encrypted file chunks stored on them. This is not an easy task and is an important layer in your data security.

If Google is automagically encrypting all our data, then what can we do to augment data security? Well, let's start with a good understanding of who holds the keys.

Controlling encryption keys

If data stored in Google Cloud is automatically encrypted, then where are the keys and how are they managed? Well, it depends. There are four main options when it comes to key storage in Google Cloud. At this point, I'm talking about the keys used for encryption, not keys used with things like service accounts and applications:

- **Google Managed Encryption Keys** (**GMEK**): Google stores and manages the keys and handles all aspects of key rotation. This is the default, and the keys are all stored in Google's internal **KMS**.

  Supported products: All

- **Customer Managed Encryption Keys** (**CMEK**) using Cloud KMS: The keys are stored in the KMS for one of your projects. You have full control over key rotation, but the key itself is still managed and stored by Google.

  Supported products: Many; for further details, see `https://cloud.google.com/kms/docs/using-other-products#cmek_integrations`.

- CMEK using Cloud **External Key Manager** (**EKM**): The keys are stored in one of the supported EKMs (Fortanix, Futurex, Thales, or Virtru). Again, you have full control over the keys and their rotation. Here, the advantage is that Google doesn't store the keys anywhere in Google Cloud.

  Supported products: Several; for further details, see `https://cloud.google.com/kms/docs/ekm#supported_services`.

- **Customer Supplied Encryption Keys** (**CSEK**): You can create and manage the keys in your on-premises environment and supply them to Google when data needs to be encrypted or decrypted. You have full control (and responsibility) over the keys and can store them where and how you like. Honestly, that's the biggest benefit and drawback of CSEK. If you don't do a good job managing and storing your keys, then you are opening your data up to a vulnerability.

  Another possible downside is that there are only two supported products: Cloud Storage and Compute Engine. For further details, see `https://cloud.google.com/docs/security/encryption/customer-supplied-encryption-keys`.

Since several of the key management options involve Google's Cloud KMS, let's take a moment to talk about some best practices.

Putting the Cloud KMS to work

Cloud KMS is a scalable, highly available, Google Cloud service for storing your cryptographic keys and performing cryptographic operations. It supports software-backed encryption keys, FIPS 140-2 Level 3 **hardware security modules (HSMs)**, customer-supplied keys, and integration with external key stores via the EKM. The service is available in every Google Cloud region, as well as in multi-regional and global configurations. Whether you use it to actually store the keys themselves or to integrate with the keys that you store somewhere else, Google's KMS can be your one-stop shop when it comes to encryption keys.

In the KMS, keys are used in the encryption/decryption process. They can be organized into logical groupings called **key rings**. Zero-to-many key rings can be placed within a project.

> **Note: KMS Isn't Designed for Application Access Keys**
>
> If you are creating an application and it needs to store a key or a credential, the KMS likely won't be a good solution, as it doesn't allow you to view or extract the keys. Bear with me for a bit, and I'll talk some about the application- and code-related best practices for GCP in the *Storing application secrets* section.

As a best practice, create your cloud KMS key rings in projects that have been built specifically for that purpose. You will likely need key projects at the environmental folder level and for specific applications and application groupings. Remember Google's Terraform blueprint for creating projects!

Once the projects are created, use IAM roles at the folder, project, key ring, and key levels to limit access. That means you are going to have to give some thought to how individual keys will need to be used, and by whom. As much as possible, group keys with similar access into key rings and/or projects to simplify IAM management.

The key IAM roles for KMS include the following:

- **Cloud KMS Admin**: Offers full access to KMS resources, but not to the encrypt and decrypt functionality. Its use should be limited to members of the security team and the Terraform service account used in the infrastructure CI/CD automation pipeline.

- **Cloud KMS Encrypter, Decrypter, and Encrypter/Decrypter**: These roles grant access to encryption or decryption for asymmetric keys, and to both for symmetric keys. These permissions can be granted as low as the key-by-key level, so be careful how access is provided. If there's a nice logical link between groups of keys and/or key rings, it can greatly simplify the assignment of permissions.

If you are planning on leveraging one of the encryption choices where you are storing keys outside of Google Cloud, make sure you plan those options appropriately. Also, remember to consider any regulatory compliance stipulations as they relate to keys and key storage.

To help manage keys, key rings, and IAM permissions related to them, Google has a KMS module in its Terraform blueprints, which you can find at `https://github.com/terraform-google-modules/terraform-google-kms`.

Excellent, now that we've discussed key management for encryption, let's examine the choices for application key storage.

Storing application secrets

I know this isn't a development class, but I at least wanted to touch on key/cred storage for applications. This is because a lot of apps need to store security and encryption keys. Application keys/creds should be considered high-security items and be handled appropriately. They absolutely should not be checked into Git repositories or embedded in code, since neither one is safe nor secure.

Google Cloud's Secret Manager is a convenient, fully managed, and easy-to-use repository for storing, managing, and accessing secrets (keys/creds) as BLOBs or text. The exact nature of the secrets is totally up to you. They could be TLS certificates, API keys, or even system passwords.

For the most part, Secret Manager should be handled similarly to the KMS:

- Create and use specific projects to hold secrets for environments and/or applications.
- Use Secret Manager IAM permissions carefully, and as usual, follow the principle of least permissions. The two main related roles are Secret Manager Admin and Secret Manager Secret Accessor.

GCP compute technologies could retrieve Secret Manager secrets from the instance metadata server. If using GKE, enable and use Workload Identity to get access to the secrets in the metadata service.

If you used the Terraform example foundation, then you might remember that it created a secrets project in the common folder, alongside the environment-specific folders, with the idea that some secrets likely belong to the entire organization and others are more stage (dev, non-prod, prod) specific.

More best practices for Secret Manager can be found at `https://cloud.google.com/secret-manager/docs/best-practices`.

> **Note: Secret Manager Limitations**
>
> While Secret Manager works well for certs, private keys, credentials, and the like, it might not be ideal for the storage of short-lived or dynamic credentials, or credentials that will scale with the numbers of users or systems.

Two other options you might consider for storing application secrets include the following:

- **Berglas (B3rg1a$)**: Berglas gives you the ability to store secrets encrypted with KMS keys in Cloud Storage. Additionally, it provides a library to integrate secret access into many GCP resources.

- **HashiCorp Vault**: Vault offers support for ACL policies. Backed by Cloud Storage or Spanner, Vault provides an excellent open source Secret Manager that can be integrated into your application.

Great! With a decent understanding of encryption and managing security keys, let's take a look at Google's **Security Command Center (SCC)**.

Improving security posture with the SCC

Google Cloud's SCC is an integrated security vulnerability and threat notification tool specifically designed for Google Cloud. Essentially, it's a detective security application that can evaluate security configurations and spot possible threats or misconfigurations, provide asset inventory information, and help spot and fix risks.

The SCC has two versions: standard (free) and premium. The standard version comes with detection capabilities for the most important threats and misconfigurations and readily integrates with BigQuery. The premium version starts with the standard feature set, and then adds a whole bunch of extra features that I'm about to tell you about.

The SCC is broken down into several major features:

- **Security Health Analytics**: This feature integrates with the Cloud Asset Inventory tool to scan assets for vulnerabilities and misconfigurations. It is part of both the standard and premium versions, though the premium version has a broader set of finding types. Additionally, the premium version includes monitoring and reporting that is specific to the CIS, PCI DSS, NIST 800-53, and ISO 27001 standards.

    Both versions support the exporting of findings to BigQuery, and the premium version adds real-time streaming of findings to Pub/Sub.

    Both versions can spot configuration errors that could prevent SCC from scanning resources effectively.

- **Event Threat Detection (premium)**: This uses ML and other technologies to analyze Cloud Logging events for malware, cryptomining, SSH attacks, IAM anomalies, and data exfiltration.

- **Container Threat Detection (premium)**: This scans containers for added binaries or libraries, malicious script executions, and reverse shells.

- **Virtual Machine Thread Detection (premium)**: This scans for cryptocurrency mining applications running in Compute Engine VMs.

- **Web Security Scanner (premium)**: This can create configurable scans of web-based applications, looking for cross-site scripting, flash injection, mixed-content, clear text passwords, insecure JavaScript libraries, and other OWASP identified threats.

To configure the SCC, start by deciding which version you are going to leverage. Many organizations stick with the standard version because it has many nice features at no extra cost. If your organization requires a more secure position, then examine both the capability set and the corresponding cost in detail: `https://cloud.google.com/security-command-center/docs/concepts-security-command-center-overview`.

Next, you will need to decide on the services you want to leverage. Service choices can be found under the **Settings** button in the upper-right corner of the SCC home page:

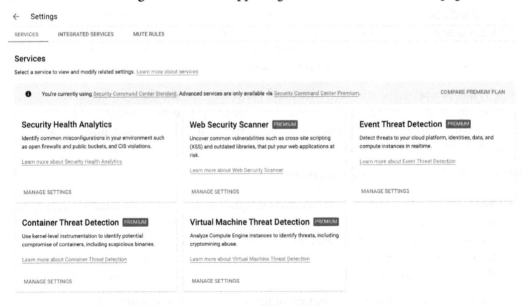

Figure 8.2 – SCC services

You will see that I'm using the standard version, so my service choices are limited. Each service will then allow you to decide what part of the organization the service should scan:

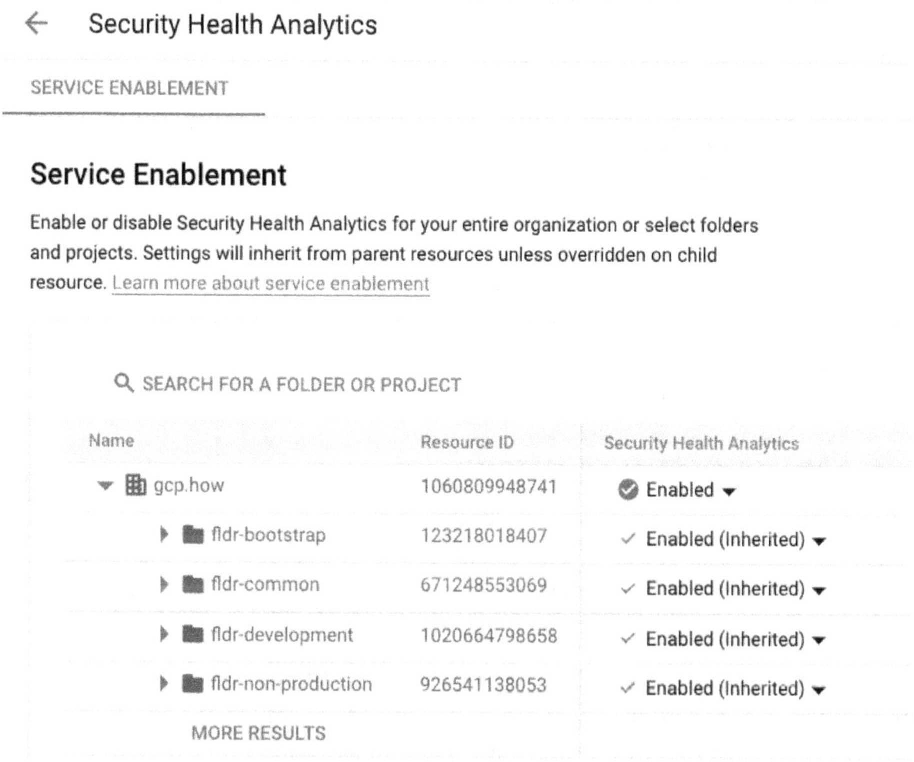

Figure 8.3 – Choosing resources to scan

The SCC creates and uses a service account as it scans. The service account name will match the format of `service-org-<org-id>@security-center-api.iam.gserviceaccount.com`, and it should be automatically granted the permissions it will need. If there are any access issues, the SCC will report them to you.

If you are using the premium version, you should create a central SCC project from which you can build and manage the Pub/Sub topic to where the automated findings messages will be sent. If you used the Terraform Example Foundation, then you might remember that it created a `prj-c-scc` project in the common folder for this express purpose.

Once the scan completes, the SCC home page provides a lot of nice summary information. The **VULNERABILITIES** tab will give you additional details, with links to specific information regarding where exactly the vulnerabilities were found:

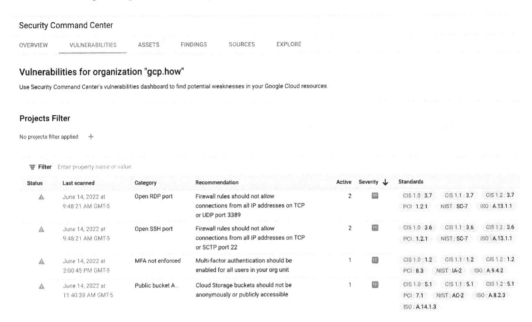

Figure 8.4 – SCC vulnerabilities list

For your foundation automation, if you are going with the premium version, make sure you create both the common/scc project (as shown in the Example Foundation) and its Pub/Sub topic. However, at this point, you will have to enable the SCC features by hand, as Terraform doesn't currently provide a resource to automate it. There is a resource that will configure the notifications (google_scc_notification_config). In the example foundation, part 1-org, you can see it being used in the scc_notification.tf file. There, they create the Pub/Sub topic, a subscription to read the findings notifications, and set up the SCC notification config. The script resembles the following:

```
resource "google_scc_notification_config" "scc_notification_
config" {
  config_id    = <notification name>
  organization = <org id>
  description  = "SCC Notification for all active findings"
  pubsub_topic = <topic name>

  streaming_config {
```

```
    filter = var.scc_notification_filter
}
```

Once you get the SCC set up and working for you, you really should enable some high-level organizational policies.

Limiting access with the Organization Policy Service

When I was in the military, I remember moving my family and myself from Aberdeen, Maryland, on the east coast, to Oceanside, California on the west. That's about as far as you can go in the continental United States without running into an ocean.
We didn't have all that much in the way of household stuff at the time, so we decided to rent a small moving truck and do the nearly 3,000-mile move ourselves. I was driving the truck and my wife was driving our car behind me.

We got everything all loaded up in Maryland and headed for the highway. Once we were through Washington DC and got out onto the open road, I started to pick up speed. Everything was normal until I got up to a little over 60 mph. All of a sudden, the truck stopped accelerating. No matter how hard I pressed the pedal, it wouldn't go any faster. At first, I couldn't figure it out. I could tell the truck had plenty of power, and I could, of course, press the pedal, but the truck just wouldn't pass 63.

Then, it hit me (figuratively speaking).

The moving company had put a speed limiter on the truck. They were fine with renting me the truck, but they didn't want me to drive it overly fast, probably for insurance reasons. In other words, I might have had the permission I needed to press the pedal, but the rental company had created a policy stating that none of their trucks could go faster than 63.

That's exactly the sort of thing you can do with org policies in Google Cloud. They provide a centrally managed way to create restrictions regarding how resources can be used, so you can set up guardrails that are designed to keep your users out of trouble and within some organizationally determined compliance boundaries.

Put simply, where IAM policies decide your access based on who you are, org policies focus on enabling/disabling features based on the resource type and where it is in your resource hierarchy.

Here is a classic example: normally, when you build a Compute Engine VM in Google Cloud, it receives both an internal and an external IP address. External IPs are nice for VMs that need to be accessible from the outside world, but they also increase the machine's surface area of attack by offering hackers a possible access point. So, the question becomes: how do you prevent machines across most or even all of your organization from having external IPs?

You can't use permissions because there is no permission on just the external IPs. You could clearly explain to everyone the importance of not using external IPs and try to ensure all of the Terraform scripts disable them, but it's still not enforceable. That would be a lot like me telling you to drive the truck under 63. Some of you would follow the rule, but for various reasons, some of you wouldn't.

Enter Google's **Organization Policy Service**.

Org policies contain one to many constraints that can be applied to a node in the resource hierarchy. Each constraint is a type of restriction, either a Boolean, enabling or disabling the resource, or it's a list of allowed or denied values.

To create and manage Google Cloud organization policies, first, you need the right permissions. The two key IAM roles related to org policies are listed as follows:

- **Organization Policy Administrator**: This is assigned at the org level and gives you full control over all aspects of org policies.

- **Organization Policy Viewer**: This is assigned at the org, folder, or project level, and it gives you the ability to view any org policies on resources within that scope.

Ah, so the ability to control org policies is an organization-level permission. Be careful who you assign it to. Typically, key security personnel and policy setters, along with your organizational administrators, will get the role.

Next, you will need to look at the available set of organizational policy constraints so that you know what you can and can't control with org policies. For a full list of constraints, please refer to https://cloud.google.com/resource-manager/docs/organization-policy/org-policy-constraints.

To help set the constraints, Google provides a Terraform org-policy resource in their blueprints (https://github.com/terraform-google-modules/terraform-google-org-policy).

For example, if I wanted to disable external IPs for all VMs in the dev folder of my org, I would need to use the vmExternalIpAccess constraint, as follows:

```
module "org_vm_external_ip_access" {
    source          = "terraform-google-modules/org-policy/google"
    organization_id = <my org id>
    folder_id       = <id for the dev folder>
    policy_for      = "folder"
    policy_type     = "list"
    enforce         = "true"
```

```
    constraint       = "constraints/compute.vmExternalIpAccess"
}
```

If I wanted, I could modify the preceding code and suggest setting the policy at the org level (`policy_for="organization"`) and using `exclude_folders` or `_projects` to make exceptions where I needed to allow external IPs.

Now, as far as recommendations for base policy constraint settings for your organization, if you look at `org_policy.tf` in the `1-org` part of the example foundation, you'll see a nice set directly from Google's security foundations guide. You might need to modify some of these for all or some of your organization:

- `compute.disableNestedVirtualization`: Use this if you want to stop VMs from supporting hardware-accelerated nested virtualization.

- `compute.disableSerialPortAccess`: Serial port access can be used to get access to VMs, typically when doing advanced troubleshooting, but it can also serve as another attack vector. This turns it off.

- `compute.disableGuestAttributesAccess`: This disables access to VM guest attributes, a special type of metadata created by VM workloads.

- `compute.vmExternalIpAccess`: Hey, we know this guy! Use this to disable all external IPs except for a defined set of exceptions. Start with a base org setting to disable external IPs for the whole organization, and then modify that with a list of exceptions as needed.

- `compute.skipDefaultNetworkCreation`: This tells Google to skip the creation of the default network when creating new projects.

- `compute.restrictXpnProjectLienRemoval`: This restricts who can remove the project lien on the Shared VPC owner project. Project liens are used to prevent accidental project deletion and must be removed before deleting.

- `sql.restrictPublicIp`: This restricts Cloud SQL database instances from having public IPs.

- `iam.allowedPolicyMemberDomains`: This restricts who can be given access to your Google Cloud environment, that is, only to the members in your Cloud Identity member domain.

- `iam.disableServiceAccountKeyCreation`: This disables the ability to create service account keys. Keyless service accounts, which only be used by GCP systems to talk to other GCP systems, are okay.

- `storage.uniformBucketLevelAccess`: This restricts Cloud Storage buckets to only use uniform (one level of permissions for all the files in the bucket) permissions.

- `iam.automaticIamGrantsForDefaultServiceAccounts`: This prevents the default Compute and App Engine service account from having any permissions.

If you are dealing with regulatory compliance or other issues that restrict you from using specific GCP regions, or GCP regions in a specific part of the world, then check out `constraints/gcp.resourceLocations`: `https://cloud.google.com/resource-manager/docs/organization-policy/defining-locations`.

Google has recently released a new org policy that can individually allow or block access on an API by API basis. Do you need to disable App Engine for your whole organization? The following link could do it: `https://cloud.google.com/resource-manager/docs/organization-policy/restricting-resources`. Note that this won't work for all resource types, so make sure that you examine the unsupported resource list.

Great! So, org policies help your security posture by blocking possible attack vectors at high levels, and they work well with some of the other features we've already discussed. Next, let's take look as some general security measures, some of which you may have already implemented.

General security elements

Mostly, these features have been discussed at various points earlier in this book, but to ensure we are all on the same (cough) page, let's review a few. As a reminder, and to save me from having to fill this section with web links, all of these topics are linked in my helpful links document at `http://gcp.help`.

There are a ton of security features related to the network, including the following:

- VPC Service Controls allow you to define security perimeters around groups of GCP services (such as BigQuery datasets, Cloud Storage buckets, and more). Make sure you have considered their application in your environment. You might recall that we previously discussed the idea of higher and lower levels of security at the network level, with the higher-level security needs adding VPC Service Controls.

- The **Identity Aware Proxy** (**IAP**) can help you secure access to web applications, specifically web applications being accessed by your employees. Essentially, it will force them to log into their GCP account before being granted access to the application. An extra feature it can provide is to give you a way to access and manage VMs without external IP addresses; it does this by working as a sort of bastion host.

- Google Cloud Armor can augment the protection provided by your HTTP(s) based applications. It works with the global load balancer to add better protection against DDoS, XSS, SQLi, and many of the OWASP top 10 vulnerabilities. Additionally, it has a WAF rules engine that you can use to configure your own allow and deny lists.

- Private Google Access could be enabled at the VPC subnet level, and it will allow resources with no external IP to access GCP services without running the traffic through the edge network.

- Cloud VPN and Interconnect can help secure traffic between your org and GCP.

- Cloud NAT can provide external access to machines with no external IP.

- Packet Mirroring can mirror network traffic to third-party SIEM tools for better network traffic analysis.

When building VM workloads, consider the following:

- Creating custom, hardened images using a CI/CD pipeline and something such as HashiCorp Packer.

- Shielded VMs provide better protection against rootkit and bootkit attacks.

- Enable OS Login to leverage IAM roles to manage SSH access. There's even an org policy you can use to enforce this: `constraints/compute.requireOsLogin`.

- Sole-tenancy for regulatory compliance when you really need exclusive access to the hardware where your VM runs.

If you work with a lot of data and you need to identify **personally identifiable information (PII)** within that data, make sure that you check out Google's Data Loss Prevention API and tools.

I could go on and on, but I think we have a good foundation of security laid, so I'm going to stop there. The very last thing I want to touch on is the tenth and final step of our foundation laying: support.

Step 10 – Setting up initial Google Cloud support

At some point, you are going to be working on a problem in Google Cloud and you are going to exhaust all your avenues of troubleshooting. It might still be an issue on your end, or it could be an actual bug in Google Cloud. When something like that happens, it's time to call in the experts.

Now, the *experts* could be a third-party support provider or a consultancy, and there are a lot you can choose from. If you're looking for that sort of help, a great place to start your search is `https://cloud.google.com/find-a-partner/`.

To get support from Google, you'll need to set up **Cloud Customer Care**.

Google offers four main Cloud Customer Care levels: basic, standard, enhanced, and premium. For a quick overview and for pricing information, please refer to `https://cloud.google.com/support`:

- Basic Support is free and comes automatically with all accounts. It includes access to Google's issue trackers to report problems (`https://cloud.google.com/support/docs/issue-trackers`) and support for billing-related issues.

- Standard Support is designed for small- to medium-sized organizations. It offers unlimited access to tech support via the Cloud Console, and critical issues are guaranteed a 4-hour initial response time during local, Monday–Friday, working hours.

- Enhanced Support is designed for medium- to large-sized companies who need faster response times and extra features. It offers 24/7 access to support cases with an initial response time of less than an hour. Additionally, it offers a lot of add-on possibilities for third-party technologies, access to account advisor services, extra support at specific times such as new releases, and support for Assured Workloads designed to help your GCP based IT services meet regional compliance regulations.

- Premium Support is Google's highest available level of support. It offers everything Enhanced Support offers with a support response time of fewer than 15 minutes. In addition, it offers Customer Aware Support to help with your onboarding into Google Cloud, Operational Health Reviews to help you measure your progress into Google Cloud, and it includes training.

To sign up for one of Google's support services, you will need to be an organizational administrator, and you'll need to reach out to Google at `https://cloud.google.com/support/docs#how_to_sign_up`.

Once your support package is in place, you can use IAM to control who within your organization has the ability to open support tickets with Google, which you can do from the **Support** section of the Google Cloud navigation menu. If I were you, I would set up a security group, perhaps `cloud-team@<your domain>`, and control your support access through those members.

Woohoo! We're almost there, peeps. Let me talk about a few final things, and we're going to have this foundation laid.

Final thoughts

You might have noticed that I didn't run through all the steps in Google's Example Foundation. If you started a foundation using Google's example, and you're still sitting with steps 0–3 done, then let me at least tell you about the steps I omitted. Step 4 helps you to set up a couple of example business units and has templates to attach them to the Shared VPCs we created in the **3-networks** step. Whether you use your bootstrap CI/CD pipeline to manage business unit-level projects is up to you. The **4-projects** step creates a new CI/CD pipeline at the business unit level, which could be a nice, more granular, approach. Then, the **5-app-infra** step builds an example application (dev, non-prod, and prod) in the business unit of your choice. Though steps 4 and 5 aren't quite as general as steps 0 to 3, they still offer nice examples of working Terraform scripts and show you a possible approach you could use in your organization.

Nice job y'all, really.

Now, if you've forgotten, in my straight job I'm a trainer. `<begin shameless plug>` I work for ROI Training `https://www.roitraining.com/`, which is Google's largest training partner, and I spend most weeks teaching people how to do things in Google Cloud. `</end shameless plug>`. I mention training because, when you move to the cloud, training is important. You likely have a lot of good people in IT, who know a lot about how your infrastructure works, but if you are going to be successful in Google Cloud, then you are going to have to create a core group of people who know Google Cloud really well. Sure, you can hire those people, but you can also put in the effort and train the best of the best of who you have now and get them up to speed on Google Cloud. You, and they, will benefit from the effort.

Besides formalized training, there's a lot of information in books, and Packt has a growing collection of Google Cloud-related titles at `https://subscription.packtpub.com/search?query=Google+Cloud`.

By now, I'm sure you know that I'm a huge fan of online documentation too, and if you have access to Google Docs, check out my helpful links file at `http://gcp.help`. If your organization doesn't like my redirect URL, or if they block access to Google Docs, then try the PDF version. It's not live, but I try to update it every few weeks: `http://files.gcp.how/links.pdf`.

Lastly, if you'd like to reach out to me, please feel free:

Patrick Haggerty, Director of Google Cloud Learning, ROI Training

Email: `patrick.haggerty@roitraining.com`

Mobile: +1-228-369-8550

Social Media: `https://www.linkedin.com/in/patrickhaggerty/`

Thanks, y'all. I hope you enjoyed the read and best of luck to you in your Google Cloud adventures!

Index

Other Books You May Enjoy

If you enjoyed this book, you may be interested in these other books by Packt:

Professional Cloud Architect Google Cloud Certification Guide – Second Edition

Konrad Cłapa and Brian Gerrard

ISBN: 978-1-80181-229-0

- Understand the benefits of being a Google Certified Professional Cloud Architect
- Find out how to enroll for the Professional Cloud Architect exam
- Master the compute options in GCP
- Explore security and networking options in GCP
- Get to grips with managing and monitoring your workloads in GCP
- Understand storage, big data, and machine learning services
- Become familiar with exam scenarios and passing strategies

Architecting Google Cloud Solutions

Victor Dantas

ISBN: 978-1-80056-330-8

- Get to grips with compute, storage, networking, data analytics, and pricing
- Discover delivery models such as IaaS, PaaS, and SaaS
- Explore the underlying technologies and economics of cloud computing
- Design for scalability, business continuity, observability, and resiliency
- Secure Google Cloud solutions and ensure compliance
- Understand operational best practices and learn how to architect a monitoring solution
- Gain insights into modern application design with Google Cloud
- Leverage big data, machine learning, and AI with Google Cloud

Packt is searching for authors like you

If you're interested in becoming an author for Packt, please visit authors. packtpub.com and apply today. We have worked with thousands of developers and tech professionals, just like you, to help them share their insight with the global tech community. You can make a general application, apply for a specific hot topic that we are recruiting an author for, or submit your own idea.

Share Your Thoughts

Now you've finished *The Ultimate Guide to Building a Google Cloud Foundation*, we'd love to hear your thoughts! Scan the QR code below to go straight to the Amazon review page for this book and share your feedback or leave a review on the site that you purchased it from.

https://packt.link/r/1803240857

Your review is important to us and the tech community and will help us make sure we're delivering excellent quality content.

www.ingramcontent.com/pod-product-compliance
Lightning Source LLC
Chambersburg PA
CBHW060526060326
40690CB00017B/3393